8
E.345
1996

OTHER TITLES OF INTEREST FROM ST. LUCIE PRESS

The 90-Day ISO 9000 Manual and Implementation Guide

The Executive Guide to Implementing Quality Systems

Focused Quality: Managing for Results

Improving Service Quality: Achieving High Performance in the Public and
Private Sectors

Introduction to Modern Statistical Quality Control and Management

ISO 9000: Implementation Guide for Small to Mid-Sized Businesses

Organization Teams: Continuous Quality Improvement

Organization Teams: Facilitator's Guide

Principles of Total Quality

Quality Improvement Handbook: Team Guide to Tools and Techniques

The Textbook of Total Quality in Healthcare

Total Quality in Higher Education

Total Quality in Managing Human Resources

Total Quality in Marketing

Total Quality in Purchasing and Supplier Management

Total Quality in Radiology: A Guide to Implementation

Total Quality in Research and Development

Total Quality Management for Custodial Operations

Total Quality Management: Text, Cases, and Readings, 2nd Edition

Total Quality Service

For more information about these titles call, fax or write:

St. Lucie Press
100 E. Linton Blvd., Suite 403B
Delray Beach, FL 33483
TEL (407) 274-9906 • FAX (407) 274-9927

S^{t}_L

Organizational Transformation and Process Reengineering

Organizational Transformation and Process Reengineering

Dr. Johnson A. Edosomwan
Group Chairman
Continuous Improvement Company
Chairman & Executive Consultant
Johnson & Johnson Associates, Inc.

S_L^t

St. Lucie Press
Delray Beach, Florida

Published by:

S$_L^t$

St. Lucie Press
100 E. Linton Blvd., Suite 403B
Delray Beach, FL 33483

QUALITY OBSERVER

The Quality Observer Corporation
Efe Quality House
3970 Chain Bridge Road
Fairfax, VA 22030

Table of Contents

PREFACE

A Wake-Up Call in a Competitive World Economy

Continuous performance improvement—the reengineering of organizational systems, structures, technology, and processes—a focus on people development, work force empowerment, product/service quality, efficiency, effectiveness, productivity, quality of management, customer satisfaction, profitability, competitiveness...these elements will remain international priorities for individuals and organizations now and in the years to come.

This is an exciting time in a world with fewer economic boundaries and increased competition in both the public and private sectors. Trends show that organizations are waking up to the challenges of the new political, economic and competitive business environments.

Many organizations are realizing that the old bureaucratic business structures can no longer respond to the increased competition and the new customer-driven market economy. Organizations are going through tremendous changes to create environments where everyone can contribute their best, where customer requirements are not only met but exceeded and where efficiency, effectiveness, productivity, quality, customer satisfaction and competitiveness are taken seriously as critical success factors.

Organizational Transformation and Process Reengineering was written to assist private and public organizations in becoming more competitive. It focuses on the critical success factors for achieving organizational transformation and process reengineering. This book provides the essential principles, tools, techniques, methodologies, models and technologies for transforming and reengineering an organization's structures, policies, procedures, processes and management systems. It evolved from several years of research, teaching, consulting and industrial work experience in helping several hundred organizations in many nations. The book was written with the belief that the right choice and approach for an organization to stay competitive is to continually transform and reengineer all aspects of its structure that can no longer respond to the current demands of the marketplace.

Organizational Transformation and Process Reengineering is organized into seven chapters and five appendices. Chapter 1 provides a description of the foundation factors and model for achieving organizational transformation and process reengineering. The core competencies of organizational transformation are addressed. The organization and process transformation model is presented; it includes the management system, the social system, the technical system and the behavioral system. The plan, approaches, deployment, evaluation and results framework for the transformation effort as well as the fundamental requirements for organizational transformation and process reengineering are presented.

The six R's (Realization, Requirements, Rethink, Redesign, Retool and Reevaluate) of organizational transformation and process reengineering are addressed. Guidelines are provided to help organizations overcome the impediments encountered in process reengineering projects.

Chapter 2 offers the principles and a step-by-step methodology for analyzing organizational structures and systems and reengineering primary, secondary and auxiliary work processes. This chapter also explains how the process reengineering methodology can be applied in both public and private organizations.

Chapter 3 provides guidelines and models for achieving organizational transformation. Specific practical ideas are presented for handling organizational transition and change management.

Chapter 4 presents reengineering process improvement models that can be utilized in both public and private organizations. Step-by-step methodologies for implementing the models are also presented. These models are commonly recommended by experts and practitioners in organizational transformation and reengineering.

Chapter 5 discusses several tools and techniques for achieving organizational transformation and process reengineering. The tools, technologies and techniques presented were guided by the error and defect prevention philosophy, customer-driven results and a people and quality first culture. The tools provide practical steps for how to map, evaluate, reengineer and improve organizational work processes.

Chapter 6 offers approaches and techniques for implementing organizational and process reengineering teams and projects; in addition, team implementation strategies are discussed.

Chapter 7 provides key survival and success factors for achieving organizational transformation and process reengineering. These factors are based on lessons learned from several public and private reengineering projects. Common implementation problems are discussed, as are strategies for overcoming them.

Organizational Transformation and Process Reengineering is designed to serve the practical needs of line executives, managers and professionals interested in the right approaches to transform and reengineer an organization for results. Advocates of efficiency, effectiveness, productivity, customer satisfaction and competitiveness will find the principles, models, methodologies, tools and techniques presented in this book very useful.

About the Author

Johnson A. Edosomwan is Group Chairman of the Continuous Improvement Company (CIC), which is an international research, consulting, training and publication organization. He is also Co-Chairman of the International Who's Who in Quality, Chairman of the Caribbean Quality Institute and Chief Executive Officer of Johnson & Johnson Associates, Inc.

Dr. Edosomwan is a senior executive consultant to more than 85 companies worldwide, as well as a past examiner for the Baldrige National Quality Award and a judge for the Sterling Quality Award and the DOE Quality Awards. He has received over 45 awards, medals and citations, including the Outstanding Young Industrial Engineer Award and Technical Innovation in Engineering, for his contributions to various fields.

The author/editor of more than 30 books and training guides, including *Customer and Market-Driven Quality Management* (Quality Press, 1994), *Integrating Productivity and Quality Management* (second edition, Marcel Dekker, 1995) and *Integrating Innovation and Technology Management* (John Wiley & Sons, 1989), Dr. Edosomwan has been credited with being the first to integrate the concept of quality and productivity and also the integration of innovation and technology management. His areas of expertise include quality, productivity, technology innovation and management, customer satisfaction, process reengineering and organization development.

He received his B.S. and M.S. degrees in Industrial Engineering from the University of Miami, his Professional Engineer degree from Columbia University and his Doctor of Science degree in Engineering Management and Economics from the George Washington University.

Dr. Edosomwan is President of the International Society of Productivity and Quality Research. He is named in several Who's Who lists, including Who's Who of Intellectuals, Who's Who in Innovation and Technology, Men of Achievement in the World and Who's Who in Distinguished Leadership, and is a recipient of the Institute of Industrial Engineers Fellow Award. Dr. Edosomwan currently serves as the Executive Editor of *The Quality Observer* International News Magazine and *Managing Technology Today*.

1

Foundation Elements for Organizational Transformation and Reengineering

N

othing is more challenging and rewarding to execute, nor more delicate to handle, nor more beneficial to implement than to transform and reengineer an organization. The organizational transformation process creates enemies, pains, change, risk, uncertainty and, most importantly, rewards that are beneficial to individuals, organizations and society at large.

Today, most organizations are facing a variety of challenges, including the need to become customer driven through competitive products and services. Many organizations are experiencing poor performance and are unable to respond to competition and customers due to outdated structures, policies, procedures, rules, tall layers of management and unresponsive management systems. Organizational transformation and process reengineering can play a key role in helping organizations gain a competitive advantage, achieve leadership in providing excellent products and services, and improve overall performance.

Reengineering the Organization

Organizational reengineering, process redesign, the scientific study of work and work measurement are not new concepts. The earlier partial version of these concepts can be traced to over 100 years of collective efforts among several academic disciplines, including engineers, management theorists, psychologists, scientists and organized labor. Exactly how organizational reengineering started is a contentious subject among researchers and practitioners. Some claim that the reengineering concept started a few years ago as a blending of several methodologies. Others claim that organizational reengineering started in the 1980s when the American auto industry, battered by Japanese rivals, began to integrate car design with assembly line automation. At that time, American automobile manufacturers embraced the ideas of just-in-time manufacturing (delivering supplies just when the factory needs them rather than warehousing inventories) and total quality management (continuously improving the quality of operations and customer service). Although problems exist in identifying the fundamental origins of process reengineering, the notion behind it is organizing, assessing, reviewing, evaluating, improving and breaking away from outdated processes that do not work.

Organizational transformation and process reengineering require eliminating old and archaic processes, policies, procedures, technologies, principles and structures that affect organizational operations. The organizational transformation and process reengineering effort can be seen as the continuous rethinking, assessment, evaluation, redesign and improvement of structures, work process elements, procedures, technologies, management systems, right-sizing and core competencies to achieve competitive performance.

As shown in Figure 1.1, the transformation process must focus on the management system, the social system, the technical system, the behavioral system and critical competitive factors.

The **management system** pertains to the way that policies,

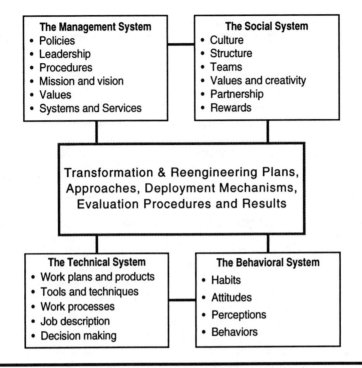

The Management System
- Policies
- Leadership
- Procedures
- Mission and vision
- Values
- Systems and Services

The Social System
- Culture
- Structure
- Teams
- Values and creativity
- Partnership
- Rewards

Transformation & Reengineering Plans, Approaches, Deployment Mechanisms, Evaluation Procedures and Results

The Technical System
- Work plans and products
- Tools and techniques
- Work processes
- Job description
- Decision making

The Behavioral System
- Habits
- Attitudes
- Perceptions
- Behaviors

Figure 1.1   Edosomwan Organizational and Process Transformation Model

procedures, practices, protocols and directives are established, enforced and maintained. The leadership system of the organization sets the tone, vision and indicators of what should be done, how it should be done and what should be accomplished. The management system carries into effect strategies, processes and project management, and it encompasses the vision, mission and values of the organization.

The **social system** has a significant impact on motivation and the ability to implement new ideas; it addresses organizational culture, structure, rewards, teamwork, values and the creativity of individuals and groups. The social system is influenced by the values of the founders, leaders, families, peers and supervisors, as well as group behaviors. An example of the cultural shift involved in a typical organizational transformation is shown in Table 1.1. Before transformation, the state of the organization is usually influenced by rigid

Table 1.1   Cultural Changes Required for Organizational Transformation

Category	Previous State	New Culture
Mission and objectives	Maximum return on investment/management by objectives Orientation toward short-term objectives and actions with limited long-term perspective	Ethical behavior and customer satisfaction Climate for continuous improvement Return on investment a performance measure Deliberate balance of long-term goals with successive short-term objectives
Customer and supplier requirements	Incomplete or ambiguous understanding of customer requirements Unidirectional relationship	Use of a systematic approach to seek out, understand and satisfy both internal and external customer requirements Practical partnership
Problem solving and improvement	Unstructured individualistic problem solving and decision making Acceptance of process variability and subsequent corrective action Assigning blame as the norm	Predominantly participative and interdisciplinary problem solving and decision making based on substantive data Understanding and continually improving the process
Jobs and people	Functional, narrow scope management controlled	Management and employee involvement, work teams, integrated functions
Management style and role of manager	Management style with uncertain objectives that instill fear of failure Plan, organize, assign, control and enforce	Open style with clear and consistent objectives, group-derived continuous improvement Communicate, consult, delegate, coach, mentor, remove barriers and establish trust
Rewards, recognition and measurement	Pay by job Few team incentives Orientation toward data gathering for problem identification	Individual and group recognition and rewards, negotiated criteria Data used to understand and continually improve processes

rules and lack of focus on customer requirements, constancy of purpose and continuous improvement.

The **technical system** includes the tools and mechanisms necessary to achieve excellent products and services. It pertains to measures which serve as the basis for improvement and planning.

The **behavioral system** relates to the fundamentals of the human side of quality, as characterized by the habits, attitudes, work patterns and behaviors of individuals and groups. Through modification of the elements in the behavioral system, it is possible to implement change that can lead to a significant breakthrough in performance. The behavioral elements are often difficult to change, and when a change is made, it positively influences the speed of organizational and process transformation.

The 6 R's of Organizational Transformation and Reengineering

The Edosomwan 6 R's, shown in Figure 1.2 (Realization, Requirements, Rethink, Redesign, Retool and Reevaluate), are recommended as a means and a language for understanding the underlying concepts involved in organizational transformation and process reengineering.

Realization

This phase of transformation and process reengineering involves recognition of the needs and challenges that face the organization and individuals. This phase requires a detailed understanding of the competitive environment, products and services, socio-cultural factors, economic and political factors, values, ethics and quality of working life issues. The realization phase should serve as a wake-up call to the organization and its individuals, bringing to their attention the fact that if an organization is to continue to operate in a competitive, global economy, it must constantly seek a means for incremental and radical improvement.

This phase also should involve the use of comparative data and information to alert decision makers and the organizational work

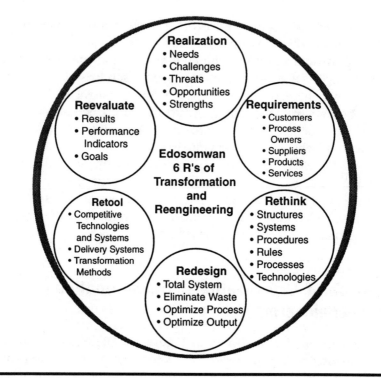

Figure 1.2 Edosomwan 6 R's of Organizational Transformation and Reengineering

force that there is a sense of urgency for incremental and radical performance improvement. The foundation premise which must be encouraged at all levels is that organizational transformation starts with individual initiative and leadership. This leadership for positive change is essential, especially at the highest level of the organization but also at every other level and in every function.

The realization message should also point out that in order for the individual or the organization to succeed, every member of the enterprise must become involved in the transformation and reengineering efforts. The challenges and problems of the present and the future must be addressed with a sense of urgency. No one should wait for the organization to shine a light toward change; rather, individuals and teams must lead and help to transform the organization through a creative and innovative process of incremental and radical problem solving. Everyone in the organization should realize that the challenge of transforming work habits, structures, systems, procedures and policies and redesigning work processes is a personal one. Radical and incremental performance improvement begins with individual commitment, dedication, ability, experience, learning, adaptation, behavior and attitude.

Requirements

Before beginning the process reengineering effort, an organization needs to define its mission, vision, values and the key requirements to satisfy and exceed customer expectations. In defining the total requirements, the voice of the customer must be heard. The requirements of the internal customer (voice of the business), suppliers and process owners as well as external customers must be captured. Following this, key product and service performance indicators and specifications are established to support the requirements. The key requirements are then used to define the ideal primary, secondary and auxiliary work processes.

Rethink

The rethinking phase of process reengineering involves reexamining all current and existing conditions of the organization. Current processes are evaluated against objectives and expected outcomes. The sources of process weaknesses and variations are identified. A careful reevaluation of outdated procedures, policies, structures, technologies, methods and work habits is performed. Any process or system that does not contribute to value and the final outcome is identified as a source of waste. In rethinking the current process, central questions such as the following are asked: Can the current process produce competitive products and services? Can the current process satisfy all the requirements of the customer and the requirements of the organization? Can the current process satisfy all of the critical performance success factors? This phase usually requires change agents and those who have the training, courage and desire to improve existing conditions.

Redesign

The process reengineering phase of redesign involves a careful understanding of the content, make-up, behavior, pattern and elements of work processes. To understand and proceed with process redesign, the organizational process must be seen as a set of logical, related tasks performed to achieve a desired outcome. A set of processes forms an organizational system. The processes are used to transform inputs into useful outputs.

A careful review of the entire system of an organization's procedures, products and services should be done. Each work element, task and job must be analyzed to determine how best to redesign the key processes in order to achieve desired outcomes. The process redesign efforts must obey three important rules.

Rule 1:   All the requirements of the organization and customers must be met.

Rule 2:   The redesign process must eliminate all sources of waste and improve the competitive position.

Rule 3:   The redesign process must fulfill job satisfaction requirements.

Process redesign can be radical, which means that all current processes are eliminated and replaced by new ones, or incremental, which means that some of the existing process elements are adapted as part of the new, redesigned process.

Retool

The retool phase involves the evaluation and adaptation of more competitive systems, such as technologies, machinery and other critical tools required to improve the production or service work processes, in order for the retooling efforts to be adequate. Current technologies are mapped according to prequalified process characteristics. Items such as mean time to failure, mean time to repair, mean time to dismantle and output per technology or machine are identified. The new set of tools is based on ideal, redesigned processes that were derived after reviewing the weaknesses of current processes and tools.

Reevaluate

The final phase of process reengineering requires the reevaluation of the entire process to ensure that once the redesign and retool efforts are completed, the new process has met its objectives. The best way to reevaluate process performance is to collect data on critical performance success factors, such as quality, productivity, customer satisfaction, market share, variation levels, profitability indexes, job satisfaction indexes and cost reduction savings. Those who perform the work have responsibility for monitoring the performance of the process because they are best qualified to control the sources of variation in the process.

Fundamentals of Process Reengineering

A coordinated, continuous improvement approach is required to rethink, redesign, retool and reinvent new processes that will perform better than existing ones. The fundamental definitions and basic elements which must be understood include the following.

The Organization Process Elements

These elements are comprised of activities and tasks. A process element may be referred to as a subprocess when it is subordinate to, but part of, a larger process. A subprocess can also be defined as a group of activities within a process which comprise a definable component.

The Organization Process

This can be defined as the organization of inputs, such as people, equipment, energy, procedures and materials, into work activities needed to produce a specified end result or output. It can also be viewed as a sequence of repeatable activities characterized as having measurable inputs, value-added activities and measurable outputs. It is a set of interrelated work activities characterized by a set of specific inputs and value-added tasks to produce a set of specific outputs.

The Organization Process Owner

The process owner coordinates the various functions and work activities at all levels of a process. The process owner has the authority or ability to make changes in the process as required and manage the process end to end, so as to ensure optimal overall performance. The process owner coordinates the inputs and outputs related to a given process.

The Organization Process Assessment and Analysis

This involves an objective assessment of how well a methodology identifies the strengths and weaknesses of the organization. The analysis involves a systematic examination of a process to establish a comprehensive understanding of the process itself. The analysis should include consideration of process simplification, eliminating unneeded or redundant elements and improving all elements involved in the process.

The Organization Process Output Measures

These are measures that pertain to how well a particular process is meeting or exceeding the requirements of self-unit internal or external customers. The organization process output measures could include, but are not limited to, quality of products and services, productivity, efficiency, effectiveness, job satisfaction, morale, revenue growth, profitability, cost of quality, cost reduction indexes, reliability of systems and technology, market share, improvement indexes and process variation indicators.

Process Management, Control and Improvement

This can be defined as the disciplined management approach of applying prevention methodologies to the implementation, improvement and change of work processes in order to achieve performance objectives, such as productivity, efficiency, effectiveness and adaptability. A critical element in the success of process management is the concept of cross-functional process focus improvement, which involves all stakeholders in achieving designed performance improvement.

Preparing the Work Force for Transformation and Reengineering

Organizational transformation and reengineering efforts bring about change. Change, regardless of the magnitude of it, results in anxiety, fear, uncertainty and hope. In order for an organization to successfully prepare the work force to accept new incremental and radical changes, it is important to know the potential sources of resistance and why people might resist the transformation efforts.

The way in which people receive a transformation and reengineering affects whether or not they feel in control of it. When we initiate a transformation effort or change, it is exciting; however, the same event is threatening when it is done to us. Most people want and need to feel in control of the events around them. Organizational change and reengineering could invoke uncertainty among those affected and also create a sense of loss of control, power, autonomy and belonging.

A second reason why people might resist even a positive organizational change is that it brings too much uncertainty. Simply not knowing enough about the next step in the incremental or radical change makes people feel uncomfortable. When people are uncertain about the beginning and end of the transformation improvement effort, they perceive danger. They feel safer clinging to the known rather than reaching out for the unknown.

A third reason why people might become uncomfortable about a potential positive transformation effort is the surprise factor. People are easily shocked by sudden changes if they are not involved and prepared. Obviously, a person's response to some radical or incremental change that is (a) new and unexpected, (b) something which he or she did not have time to prepare for mentally or (c) something in which he or she was not invited to participate is either total withdrawal or defensive resistance.

The fourth reason why people will resist change is that change requires people to become conscious of and to question familiar routines and habits. If accepting a change means admitting that the way things were done in the past was wrong, people are certain to resist. Nobody likes losing face or feeling embarrassed in front of peers. Sometimes making a commitment to a new procedure, product or program carries with it the implicit assumption that the "old ways" are bad. This puts individuals in the uncomfortable position of either looking stupid for their past actions or being forced to defend them; thus they would argue against any change.

Sometimes people resist change because of personal concerns about their future ability to be effective after the change. Can I do it? How will I do it? Will I make it under the new condition? Do I have the skills to operate in a new way? These concerns may not be expressed verbally, but they can result in finding many reasons why change should be avoided.

People may resist change for reasons connected to their own activities. Change sometimes disrupts other plans, projects or even personal and family activities that have nothing to do with the job. The anticipation of those disruptions causes resistance to change.

One reasonable source of resistance to change is that change requires more energy, more time and greater mental preoccupation. Past negative experiences—those cobwebs of the past that get in the way of the future—cause people to resist change; negative experiences, however, are a reality in organizational life. Anyone who has ever had a gripe against an organization is likely to resist when the organization tells him or her to do something new.

The last reason why people resist change, which in many ways is the most reasonable of all the reasons, is the real threat posed by change. Sometimes change creates other changes, even in winners. Change is never entirely negative. It is also a tremendous opportunity, but even in that opportunity there is some small loss. It can be a loss of the past; a loss of routines, comforts or traditions; or, more importantly, a loss of relationships that have become close over time. Things will not, in fact, be the same anymore, which is why every change must be named and implemented positively. The following steps are recommended to prepare the organization team for transformation and reengineering.

Step 1: People Involvement

In beginning the organizational transformation and reengineering effort, it is important to identify the critical voices (such as customers, unions, suppliers, process owners, managers and employees) and involve them in the process of diagnosing for change. In diagnosing for change, employees should understand the business thoroughly and find out what is happening, what is likely to happen in the future and how the anticipated changes will affect their own organization. Specific attention should be paid to market-driven changes, customer expectations, technology changes, skill mix changes, product development cycles, regulations, competitors, cultural changes and service and manufacturing capability, among others.

Step 2: Planning and Preparation

Most change effort begins with the identification of a problem or stems from a need presented by a new market requirement. Change efforts involve attempting to reduce discrepancies between the real and the ideal. It should be pointed out that reducing discrepancies between the actual and the ideal means thinking the change through thoroughly and carefully. In this step, the participative management style works very well. Planning and preparation include the following:

- Description of the change and communication of the anticipated benefits.

- Obtaining input from those who will be affected by the change, those who will help to implement the change and those who will benefit from the change. Input from customers, employees, peers, superiors and subordinates should be encouraged.

- Assess the organizational readiness to make the change. This usually requires answering the following questions:

  o What is the maturity level of the people involved?

o   Are they willing and able to make the change?

o   What leadership, decision-making and problem-solving skills are available, and what are the assumptions behind the change?

o   What are the expected risks and benefits?

o   Is everyone ready to undertake the change?

- Prepare the change plan with different options and highlight the preferred plan and timetable.

- Anticipate the skills and knowledge required to master and implement the change.

- Focus on the changes that are critical for success. Change the most important things once, if possible, to encourage stability, and change routine and minor operations when appropriate.

Step 3: Develop Change Agents and Transition Structures

Change agents must have the skills and ability to diagnose a given situation and develop acceptable solutions. The essential skills and abilities include, but are not limited to, the following: facilitating, listening, participative designer, team leader, catalyst and courage. The wisdom to know when to push the change versus when to step back and let people accept the change over time is also required. Humility can easily facilitate the implementation of change. Successful change management cannot be achieved without the proper communication channels. Appropriate new communication channels are required to get people involved and to let them know why the change makes sense. When appropriate, create a transition team to oversee the change and develop policies and procedures that make implementation of the change easy.

Step 4: Change Execution and Implementation

Most change fails to yield expected results, not because it is not good but because the ingredients and mechanisms for implementation were not executed properly. Leadership, vision, courage, empathy, humility and wisdom are required to implement transformation and reengineering projects. The work force must be inspired and motivated to change. Everyone within the organization must be helped to understand the environment and how to implement the desired change. Everyone in the organization needs to know how to implement the aspects of the change that affect their own work. In addition, the following must be done to ensure effective implementation:

- Help the managers and the members of the work force understand how to define the change and how to understand the management, communication and training process involved in managing the change.

- Develop the managerial tools and skills required to manage the change. These include analytical, behavioral and organizational management skills; willingness to change; commitment to change and the desire to manage the change.

- Recognize the informal organization and provide a positive climate for those affected to respond and accept the change quickly. An organization can be viewed as a social system consisting of a loose network of small groups of people. People in these groups can form a strong bond of loyalty to each other. These groups should be used to institute the change. If the informal groups and the informal leaders accept the proposed change, the change will occur much more smoothly. If they oppose the change, it may be nearly impossible to implement. Therefore, it is important to identify and get the informal groups involved in the change execution and implementation.

- Develop "change agents" for change execution and effective implementation. It is very important for the leader of the change to seek the active support of the critical mass of the work force. The critical mass usually represents a sufficient number of influential people who support the proposed change. When the critical mass support is obtained, the change execution and implementation occur much more smoothly.

- Assess the readiness of the enterprise to make the change, allowing enough flexibility for people to prepare for the change and deal with the consequences. Plans should be in place to deal with the logical, rational and irrational sides of change.

In order to implement a change successfully at any level of the organization, the following steps are suggested:

Step 1: Assess the magnitude of the proposed change and the vision for it.

Step 2: Divide the change into manageable and familiar steps and specify implementation timing.

Step 3: Communicate the change effectively.

Step 4: Assess the potential response to the change. Identify potential avid supporters, those who are undecided and the resisters.

Step 5: Develop and implement strategies to win the support of everyone, including the undecided and resisters.

Step 6: Develop and implement actions to assist those affected by the change and follow up on open issues.

Summary

Organizational transformation and reengineering can be initiated through (a) leadership vision; (b) pressure from competition; (c) input from customers, suppliers, process owners, unions and employees; (d) business crisis due to poor performance and (e) third-party assessment of organization strengths and weaknesses. The foundation elements discussed in this chapter must address the management system, the social system, the technical system and the behavioral system. An organization interested in reengineering all of its elements must go through the process of the 6 R's: Realization, Requirements, Rethinking, Redesigning, Retooling, and Reevaluation of the total organizational elements. The key to a successful transformation and reengineering effort is involving the people and managing the change process so as to minimize resistance, pain and anxiety while maximizing the benefits of improved productivity, quality, customer satisfaction, efficiency, effectiveness, profitability and competitiveness.

2

Principles and Methodology for Organizational Transformation and Reengineering

---

Organizations and people are led by vision and guided by mission and purpose. The organizational continuous improvement process must be guided by results-oriented principles and methodology.

Guiding Principles for Transformation and Reengineering

Based on practical work experience in both the public and private sectors, the following Edosomwan performance improvement principles are recommended for a successful organizational transformation and reengineering effort.

Principle 1: Leadership and Work Force Focus on Constancy of Purpose and Continuous Improvement

The organization leaders should create a sense of urgency for continuous improvement and constancy of purpose. Providing constancy of purpose involves a focus on essential management practices, including goal setting, planning, policy deployment and efficient implementation of improvement projects. Planning for transformation and reengineering is accomplished through a structured and methodical approach which links together the diverse activities and levels of the organization. Strategic, tactical and operational transformation plans must address both the needs and means to achieve meaningful results.

Planning is an absolutely essential element of developing the right structure, policies, procedures, training, teamwork, systems and overall framework for continuous improvement. It is important that the voices of the customers, suppliers, process owners, unions, management and work force be heard and incorporated in the transformation plans. Appropriate goal setting must be linked to the strategic, tactical and operational planning process by a demonstrated management commitment to specific means and resources for achieving the specific objectives and goals articulated in the plans. Specific measurable goals and objectives should be set at all levels. The goals must reflect the requirements of the internal and external customers as well as the overall vision provided by management and the work force.

Once the goals are set, the next step involves policy deployment, which requires internalizing and communicating transformation and reengineering objectives throughout the organization. This process is a cascade of objectives from the higher levels to the lower levels of the organization. A network of interlocking transformation and improvement teams should be used to translate higher goals and strategies into meaningful actions. The policy deployment process also should be used to set priorities, including a focus on the following tasks:

- Promote the transformation and reengineering vision, policies, goals, objectives and expected benefits.

- Ensure that comprehensive plans are in place at all levels to manage transformation projects.

- Develop a comprehensive guideline for reengineering and restructuring work processes.

- Provide training to everyone on the skills required to achieve successful transformation.

The efficient implementation of the transformation and reengineering outcomes should be the responsibility of the process owners and the individuals assigned to specific implementation tasks. The implementation efforts should focus on achieving the performance objectives, assessing outcomes through use of measures, optimizing resources and providing an environment and culture that support expected results.

Principle 2: Simplify Structures, Processes, Procedures, Policies, Systems and Programs

Organizational and process reengineering efforts should focus on reducing layers of management and structures that cause delays. Processes that have too many steps should be redesigned to achieve few steps and improvement in cycle time. Outdated procedures, policies, systems and programs need to be evaluated, simplified or eliminated when they no longer serve any useful purpose.

Principle 3: Eliminate and Minimize Waste

Every member of the organization should take it as a personal responsibility to eliminate waste. The purchase of goods and services should be determined based on what is actually needed instead of what is nice to have. Wasted personal time and company resources as well as any errors that cause rework should be eliminated.

Any activity, task, work elements system or process that does not add value to the final output or outcome desired by an organization

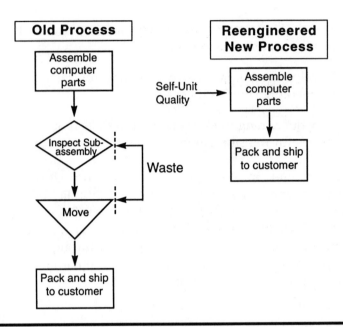

Figure 2.1   Simplified Example of Waste Elimination and Reengineered Process

or individuals can be defined as waste. Waste results in delays, increased costs, increased cycle time, decreased productivity and an increase in rework. Excessive transportation, storage steps and duplicate machines, such as computers without full utilization, can be defined as waste. Waste can be eliminated by learning to document process steps and evaluating them for waste sources and sequences. An example of a process that was reengineered in a computer company is illustrated in Figure 2.1.

Principle 4: Design and Implement Parallel Processes

Linear processes produce long cycle times. The linear process requires that everything must wait for the completion of a previous step before the next step can begin. As shown in Figure 2.2, parallel processes can easily be achieved through process resequencing and rearrangement. The parallel process reduces cycle time, process bottlenecks and delays that affect customer satisfaction.

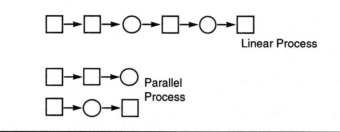

Figure 2.2 Linear and Parallel Processes

Principle 5: Focus on Constant Innovations and Use of Technology to Improve Processes

Constant innovation of work processes can lead to breakthrough and improvement. Everyone should be encouraged to examine their own job and work habits to find better ways to creatively accomplish the same work. Technology should be utilized to increase productivity and improve quality when the benefits far outweigh the costs. However, the social aspects and impact of new technology should be analyzed thoroughly before introduction to the work environment.

Principle 6: Create and Implement Performance-Based Measures to Assess Process Outcomes

If the performance of any system or process is not measured, it cannot be meaningfully improved. Quantitative and qualitative measures that address both objective and subjective elements should be implemented. The measures recommended should include, but are not limited to, the following:

- Revenue per employee

- Percent defectivity

- Process control limits

- Cost of quality

- Total factor and partial productivities

- Profitability ratios

- Customer satisfaction indexes

- Reliability rates

- Morale and job satisfaction indexes

Principle 7: Implement Error and Defect Prevention Philosophy at All Levels

Error and defect prevention mechanisms are the primary source of controlling process variation. Everyone should be trained to prevent errors and defects at the source. Processes that have no errors and defects produce error- and defect-free outputs. An error and defect prevention philosophy can best be implemented when the following mechanisms are in place:

- Appropriate tools and techniques for solving problems

- Adequate training provided to handle the challenges and process problems

- Decision-making latitude and constructive empowerment provided to people to handle their own responsibilities effectively

- Existence of a support structure for implementing new process improvement ideas

Several types of errors and defects can occur due to technology, methods, data, process complexity, training, techniques, materials, machines, manpower, environmental factors, design flaws, policies, procedures, decision-making process and lack of experience and attention to detail. Errors and defects can be prevented and corrected by reviewing process and performance data, evaluating the skill base, examining causes and effects of problems, and implementing variation control mechanisms within the process.

Principle 8: Define Process Owner(s), Stakeholders and Suppliers

No process can be fully reengineered without defining the key owners. A process owner is the person responsible for overseeing the evaluation, assessment and implementation of process improvement ideas. When there is no process ownership, the responsibility for reengineered, new ideas has no home. In order to avoid this, primary, secondary and auxiliary process owners must be defined at all levels. The definition of the process owners should also include key process outcomes, measures, suppliers, process customers, process dependencies, decision-making gates and problem resolution sources.

Principle 9: Involve Customers, Process Owners, Suppliers and Unions in Reengineering Efforts

In order for a process reengineering effort to be successful, the voices of the customers, process owners and suppliers must be included in the reengineering effort. Understanding and responding to the needs, inputs and expectations of these voices are essential in order to identify potential beneficial areas for reengineering as well as identify sources of implementation obstacles to new ideas. The dialogue between these voices also ensures that no inputs, process variables or outputs are ignored in the process reengineering effort. In order to achieve the full involvement process, the reengineering team must take the following actions:

- Provide a framework for defining and identifying process owners, customers, suppliers and unions

- Provide policy that requires each individual and organizational work unit to know its suppliers, process owners, customers and unions

- Define a measurement system that evaluates total performance and the relationship between process owners, customers, suppliers and unions

Three variables link suppliers, process owners and customers together: measures of performance, common missions and outputs, and common challenges and problems. When these three variables are analyzed adequately, it is not difficult to recognize the significance of involving suppliers, process owners and customers in reengineering improvement projects.

Principle 10: Promote Radical and Incremental Improvements

Some processes are so outdated in their ability to deliver useful outputs that they should be radically reengineered. The radical redesign of a process involves the complete elimination of all process steps and elements and the substitution of a complete new process package. This approach is only cost effective and beneficial if the benefits of the radical new process clearly outweigh the costs of maintaining and improving the old process. Incremental process reengineering calls for a careful approach to promote improvements; a systematic approach is needed for the identification of small improvement opportunities in existing structures, processes, tools and systems in order to incrementally change the performance outcomes.

An incremental reengineering effort is more acceptable in many environments because it allows for a positive approach in the management of the transition elements and the changes involved in the reengineering effort. Incremental reengineering allows low, moderate and high risk takers to work together in addressing simplification of organizational structures, processes and systems in order to make them more efficient, safer and productive, as well as less costly.

Transformation and Reengineering Methodology

Organizational transformation and reengineering cannot be achieved in a crisis management manner. All aspects of the transformation and reengineering effort must be coordinated properly. The final outcomes can be achieved if the ten-phase methodology, illustrated in Figure 2.3, is utilized. This methodology has been implemented in more than fifty organizations.

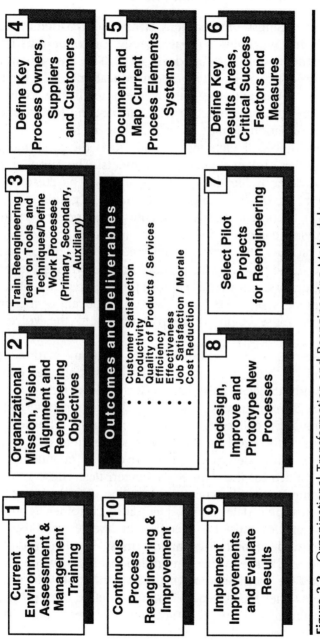

Figure 2.3 Organizational Transformation and Reengineering Methodology

The following content appears within the figure:

1 Current Environment Assessment & Management Training

2 Organizational Mission, Vision Alignment and Reengineering Objectives

3 Train Reengineering Team on Tools and Techniques/Define Work Processes (Primary, Secondary, Auxiliary)

4 Define Key Process Owners, Suppliers and Customers

5 Document and Map Current Process Elements / Systems

6 Define Key Results Areas, Critical Success Factors and Measures

7 Select Pilot Projects for Reengineering

8 Redesign, Improve and Prototype New Processes

9 Implement Improvements and Evaluate Results

10 Continuous Process Reengineering & Improvement

Outcomes and Deliverables
- Customer Satisfaction
- Productivity
- Quality of Products / Services
- Efficiency
- Effectiveness
- Job Satisfaction / Morale
- Cost Reduction

Phase 1: Current Environment Assessment

The current business environment is assessed to determine areas of strength and weakness so that strategy for continuous improvement implementation can be developed. The assessment of the operational environment serves as an educational step for the top management team and the improvement teams by defining the current state. The assessment provides a baseline for product, process and information quality, as well as human resource utilization. One of the important elements of success is the total understanding of supplier and customer requirements. The supplier plays a key role in determining the quality of inputs. The customer is the final judge of the quality and competitiveness of goods and services. The goal is to find methods of identifying supplier requirements and identifying customer needs, wants and desires and translating them into product and service requirements. Those who create the output and those who receive it should agree on clear, measurable goals for the work. In order to capture the requirements of suppliers and customers in products and services, producers of the products and services must begin to listen carefully to the voice of the customer.

The assessment process begins with obtaining marketing information and customer perception data on existing and competing products and services. This data collection provides the baseline required to develop a clear understanding of customer demands, and data must be collected on an ongoing basis. The next step is to break down the product and service system into part or process characteristics and determine qualitative and quantitative issues and target values for the defined characteristics. These values must then be compared to the customer baseline data. In order to compare the customer data to product data, customer requirements and characteristics must be specified in like terms. Customer requirements and market information must be analyzed and the data synthesized for relevance. Interrelationships between characteristics should be considered at this time. Once this critical analysis is complete, the final step is to fully define the product and service characteristics and technical measurements that meet customer requirements.

The assessment of the business environment must be done continuously. This requires an ongoing partnership with suppliers and customers. Such a partnership provides the basis for total understanding of market requirements, needs and wants. The following suggestions are recommended for enhancing the partnership among the supplier, process owner and customer. First, maintain continuous communication with customers and suppliers. This requires one-on-one contact, an on-line communication channel, telephone contact and periodic site visits. Also, perform periodic surveys to access the degree of customer satisfaction and use the surveys as a feedback mechanism for understanding areas for improvement. Provide feedback to customers on improvements achieved in the product and service offerings. Second, encourage customer participation in developing quality excellence strategies for new products and services. Encourage suppliers to participate in implementing a quality excellence and total customer satisfaction program. This can be very useful in situations where the suppliers' commodities are essential to the products and services delivered.

The assessment of the current business environment must be comprehensive. The assessment should provide a baseline for human resource utilization and products and processes. The major areas to focus on are as follows:

Organizational Structure: The effectiveness of the organizational structure, span of control, policies, procedures and decision-making processes.

Leadership for Transformation and Reengineering: The senior management vision and commitment to organizational transformation and improvement.

Culture and Environment: The culture that promotes continuous improvement and building blocks and values that guide the organization toward excellence. This also includes the operating philosophy at all levels of the organization, including work force empowerment, ability to encourage partnership for progress at all levels and a customer-driven culture of excellence.

Information Utilization and Analysis: The effectiveness of the company's collection and analysis of quality improvement and planning.

Strategic Quality Planning: The effectiveness of the company's integration of the customers' quality requirements into its business plan.

Human Resource Utilization: The success of the company's efforts to realize the full potential of the work force for quality management.

Quality Assurance Results: The effectiveness of the company's systems for assuring quality control in all its operations and in integrating control with continuous quality improvement. The demonstration of quality excellence is based upon quantitative measures and results.

Customer Satisfaction: The effectiveness of the company's utilization and integration of new technology into new and mature products and processes.

Innovation, Technology and Process Management: The ability and commitment of the company to encourage and manage innovation at all levels of the organization.

Supplier Management: The successes of the company's efforts to encourage and develop a supplier network that utilizes effective quality assurance and management techniques and controls.

A review of the above areas provides the basis for understanding the organization's culture, its technical system, the management philosophy, its products and services, and its ability and commitment to meet customer requirements. The outcome of the organizational assessment should be an aggregated analysis of a series of surveys, interviews and data collection efforts. It should conclude with prioritized recommendations to target the improvement strategy. Senior management and staff also need to be trained in transformation goals and change management, as well as how to deploy the reengineering objectives and goals.

Phase 2: Define Organizational Mission, Vision, Values and Process Objectives

This phase involves defining the mission and vision of the organization. The mission should reflect the nature of business, products and services, and the vision should specify the expected quantum desired improvement. The process objective should be tied to mission and vision. The senior management of the organization has the ultimate responsibility to define a clear and shared vision of where the organization is headed. It is very important that the vision is written, disseminated and understood by everyone. A vision is only useful if those who will implement it know about it and are guided by it. Appropriate transformation cannot happen unless values are established. The organization's values should extend beyond the basic requirements for legal behavior. The values should reflect fundamental beliefs and expectations to every employee through a clear statement and enforced actions. These values should include, but are not limited to, customer satisfaction priorities, honesty, fairness, teamwork, strength, business ethics and integrity, and commitment to growth.

Phase 3: Train Reengineering Team and Define Primary, Secondary and Auxiliary Work Processes

The creation of an organizational transformation and reengineering team is an excellent way to achieve results quickly. The team must be trained in change management, cohesiveness, tools, techniques for problem solving and decision making, measures and critical success factors, people sensitivity and project management. Once the team is trained, the various levels of the process should be identified. The following are three recommended levels of processes:

Primary Processes: Primary processes are the key processes within an organization which produce the main products, services or final output for the external customer.

Secondary Processes: These are the second-level work processes within an organization. They support the key processes to deliver the final output to the external customer.

Auxiliary Work Processes: These are the third-level work processes which support the secondary work processes to deliver the final output.

The process boundaries should also be documented, including the beginning and end points. The relationship between the beginning and end point of each primary, secondary and auxiliary process must be established.

Phase 4: Define Key Process Owners, Suppliers and Customers

The Edosomwan tri-level definition of processes, suppliers, process owners and customers, shown in Figure 2.4, represents an organization as a total system, which contains inputs, processes and outputs. The system is maintained by suppliers, process owners and customers. Any transformation and reengineering effort must understand these relations. The central driving forces in successful transformation are the relationship, ownership and accountability of suppliers, process owners and customers. In order to adequately define these terms and relationships, the following Edosomwan definitions are needed:

External Customer: The final arbiter who receives the work output. The external customer is the judge of whether or not the quality, price, delivery, schedule, service and other specifications are exceeded and met.

Internal Customer: The internal judge of the output that comes from another department or individual.

Self-Unit Customer: Every individual is a self-unit customer of themselves. Excellence at the individual level calls for self-inspection, a disciplined attitude, better work habits, self-measurement and evaluation, and self-ownership and accountability for the final output.

Supplier: Anyone responsible for supplying inputs to a process or system. There are also three levels of suppliers: (a) external suppli-

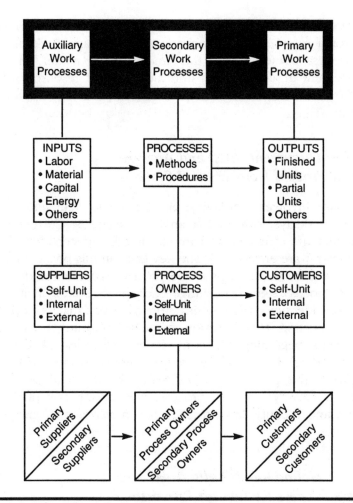

Figure 2.4 The Edosomwan Tri-Level Definition of Processes, Suppliers, Process Owners and Customers

ers, who provide desired inputs from outside the organization (also called external vendors); (b) internal suppliers, who provide input from one department or one individual to another within the same organization and (c) self-unit suppliers, which means every individual supplies himself or herself with critical input, such as personal time and materials.

Process Owner: The person charged with the ultimate responsibility for a set of primary, secondary or auxiliary work processes. The process owner is also the final decision maker on process improvement implementation and allocation of resources and can take the blame or credit for the problems or progress involved in the process.

Phase 5: Document, Map and Analyze
Current Process Elements and Systems

This phase involves identifying process steps, elements, activities, sources of delays, bottlenecks and inefficiencies. Current process parameters and characteristics are baselined, and steps for overcoming problems are envisioned. The key tasks in this phase include (a) identify and record all steps in the process; (b) detailed description of each process step; (c) arrange each process step in its correct order; (d) identify each process step by type, such as operation, delay, inspection, transportation and decision gates; (e) determine the process beginning and end; (f) identify process inputs and outputs and (g) determine the process metrics. At the completion of this phase, the process analysis team or individual should gain general familiarity with the process, identify the key purpose for analysis and identify opportunities for improvements and current strengths.

Phase 6: Define Key Results Areas and
Critical Success Factors for Processes

In order for each major primary process to be redesigned, key results areas and critical success factors should be identified. The success factors include, but are not limited to, the following: customer satisfaction, financial measures, productivity measures, quality measures, job satisfaction measures and profitability measures. It is strongly recommended that critical measures be monitored periodically. When a new reengineering project is started, the measures before and after should be documented. This will enable a positive comparison of whether or not progress has been achieved through the reengineering efforts.

Phase 7: Select Pilot Projects for Reengineering that Focus on Customer-Driven Results

This phase involves selecting pilot projects that are good candidates for improvement. The goal of process reengineering is to eliminate waste, delays, redundancies and excessive bottlenecks. The elements to consider in selecting pilot projects should, therefore, focus on opportunities for improving the following: (a) inefficiency in process steps; (b) time-consuming delay steps; (c) redundant inspection steps; (d) sources of rework and repairs; (e) excessive paperwork, policies and procedures; (f) sources of lost productivity and efficiency and (g) excessive process steps that increase cost of quality.

Phase 8: Design, Improve and Prototype New Processes

This phase involves bringing those involved in the process to a brainstorming session to identify new initiatives to improve processes and the entire organization's systems. One way to ensure that reengineering has a cross-functional perspective is to assemble a team that represents the functional units involved in the process being reengineered and all the units that depend on it. Rather than looking for opportunities to improve the current process, the team should determine which of its processes really add value and search for new ways to achieve the end result. When looking for improvement opportunities, ideas to consider should include, but not be limited to, the following:

1. Questioning the value of every process step

2. Eliminating non-value-added work elements and activities

3. Reducing process complexity

4. Using technology or systems to improve performance

5. Combining process steps through linear or parallel design

6. Selecting the most efficient method of transportation, movement and sequencing

The key to redesigning and improving an old process is to have detailed knowledge of what is done today, how it is done and how it should be done in the future. The appropriate rule to follow is "evaluate, minimize, simplify, combine or eliminate" all sources of waste, redundancies, delays, transportation and non-value-added steps.

Phase 9: Implement Improvement and Reevaluate Processes for Continuous Improvement

The key to achieving radical quantum performance and results is implementing improvement ideas at all levels. Improvements often reflect drastic changes in culture, structure, systems, policies, procedures, technologies and tasks. The improvement may also reflect significant changes in existing jobs by integrating task elements and empowering workers with the authority to deliver better results. All process improvement ideas must be reevaluated periodically to ensure that the expected results are achieved and the desired outcomes are obtained. Process improvements can be implemented in four ways:

1. Radical switch to the new method

2. Incremental or gradual switch to the new method

3. Controlled implementation of portions of recommended improvements

4. Pilot testing of total improvements

The pilot testing approach usually produces the best results because it allows for adjustment and correction of any potential problems. Once improvements have been implemented, they must also be monitored to ensure that the expected benefits are realized.

Phase 10: Continuous Reengineering and Performance Improvement

Reengineering organizational work processes for performance improvement is not a one-time effort. It involves an ongoing process of understanding the current process, task boundaries and all variables within the environment. The vision for improvement should be continuous. Everyone within the organization needs to be involved in the ongoing identification of new improvement opportunities; as for the available opportunities, they should focus on the projects that will yield the greatest benefit. Ongoing improvement cannot be realized unless the scope of the improvement effort, responsibilities and ownership of results are clearly defined. It is recommended that a breakthrough in new knowledge and the pilot approach to testing should be encouraged. In order to avoid a sudden stop in process improvement progress, strategies for overcoming resistance to change must be implemented. Continuous measurement, evaluation, planning and improvement of process improvement ideas are recommended.

Summary

Once an organization has decided to embark on a transformation and reengineering effort, it must follow the results-driven performance improvement principles and methodology. The performance improvement principles, combined with the results of current environment assessment, should be utilized to improve primary, secondary and auxiliary work processes. The results of the transformation and reengineering efforts cannot be achieved without the appropriate involvement of the stakeholders, members of the work force, suppliers, process owners and customers. The message that should be communicated at all levels is that as long as the organization exists and people have jobs, an urgency exists which requires everyone to focus on the constancy of purpose and the continuous improvement of all the elements of work, products and services.

3 Organizational Transformation Guidelines and Models

E very organization can benefit from *continuous improvement through new ideas. The shape, size, growth and success of the organization of the future depends on how well the organization of today is transformed to deal with current and future challenges.*

Practical guidelines and models for transforming organizational structures, systems, processes, procedures and performance are offered in this chapter. Some of the models presented are adapted from the works of other successful authors and implementors of organizational performance improvement models.

Organizational Transformation Guidelines

It is important to have guidelines for organizational transformation because they facilitate integration, execution and transition from the current state to the future organizational state in the most effective manner. The following guidelines are recommended for handling the organizational transformation process.

1. Obtain Commitment to the Transformation Effort

Perform a thorough evaluation to understand the organization's climate and culture. Secure and obtain total support of everyone who will be affected by the transformation and changes. The process of building commitment begins with the senior management of the organization and is continued until all the stakeholders and members of the work force are committed. Figure 3.1 provides the Edosomwan description of the various stages involved in obtaining commitment to change and the transformation effort. Four major stages are involved as people move through the change acceptance process.

Stage 1. Visualization and Awareness: People have a sense of what needs to be done differently. They can visualize the elements of the future, but they are uncertain how the vision for transformation can be realized.

Stage 2. Apprehension and Self-Concern: The means for realizing the transformation effort becomes very clear, and people are concerned about how they will be affected by the transformation and changes. Several personal questions are asked: How will the changes affect careers?...compensation?...job security?...decision-making latitude?...the quality of work life?

Stage 3. Reality and Mental Tryout: People now view the transformation effort as reality and know that changes are inevitable. People's attitudes begin to shift to either avid supporters or resisters. The supporters will find ways to make the perceived changes work for them, and the resisters will find ways to defeat the new ideas. At this point, most resisters are converted to supporters if the benefits of the changes are explained properly.

Stage 4. Transformation and Change Acceptance: People accept the changes and are willing to deal with the consequences. People will accept changes if they perceive there are good reasons for them. Stage 4 becomes easier to achieve if there are specific

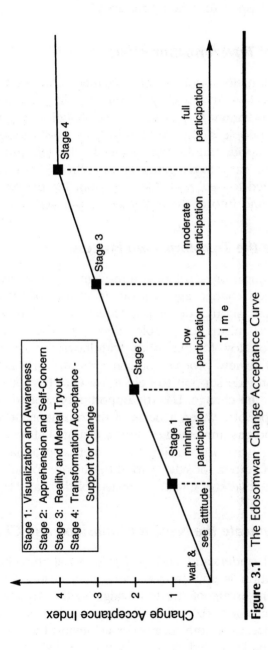

Stage 1: Visualization and Awareness
Stage 2: Apprehension and Self-Concern
Stage 3: Reality and Mental Tryout
Stage 4: Transformation Acceptance -
          Support for Change

Figure 3.1 The Edosomwan Change Acceptance Curve

implementation steps for the change and the timing and content are realistic and open to new opportunity.

2. Plan the Transformation Effort

Any transformation effort is about change. Change, no matter how small, yields fear and anxiety. In order to improve the acceptance level, it is important to plan the implementation activities ahead of time. Give people a chance to step back and understand what is occurring. Provide training for people at all levels and use transition management teams to mitigate concerns. Provide counseling and the skills required to deal with the transition. Let people know where they stand and inform them of their responsibilities.

3. Manage the Transformation Process

The transition in organizational transformation can best be handled if a respected change agent is put in charge of the process. The change agent should develop adequate transition structures and support mechanisms for people affected by the change. It is also important to allow for temporary withdrawal by individuals. Establish a clear review process for concerns and grievances that may arise and provide support to help those who have special difficulty adjusting to the change. Use the opportunity created by the change to reward people. When aspects of the transition become difficult, find out who has implemented such a change before and learn from their experience. Use outside help when appropriate, and implement a continuous, positive support structure to help people deal with the positive and negative impacts of the transformation effort.

4. Communicate the Need for Transition and Change

People will understand and deal with what they know. Positive communication is an essential element in ensuring that people at all levels are aware of the following: (a) the need for transformation and changes, (b) the goals and objectives involved in the change process, (c) the means for achieving the changes, (d) the benefit of the expected changes and (e) individual and team roles and responsibilities.

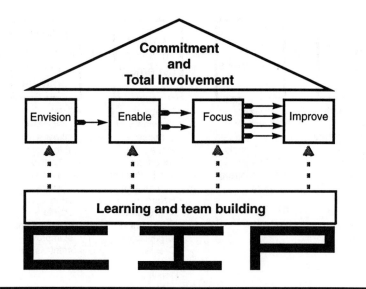

Figure 3.2   LMI CIP Transformation Model

LMI CIP Transformation Model*

The LMI CIP Transformation Model, shown in Figure 3.2, focuses on the organizational and behavioral changes needed to instill and sustain a culture of continuous improvement in organizations.

The organization develops a unified, consistent vision of its goals and objectives and achieves that vision by providing the leadership and resources necessary to implement changes, as well as eliminating transformation barriers. Broad goals are focused down through all the organization's layers, and improvement practices follow a structured, disciplined methodology. Training and team building have fundamental, supporting roles throughout the LMI CIP Transformation Model, as people and groups in the organization must be trained in appropriate subjects at the appropriate times and groups must learn to function as teams. The ultimate objectives are to (1) establish a perpetual and total commitment to continuous improvement throughout the organization and (2) involve everyone. Con-

---

* The source of the LMI CIP Transformation Model is *An Introduction to the Continuous Improvement Process: Principles and Practices* by Brian E. Mansir and Nicholar R. Schacht, Logistics Management Institute (1989).

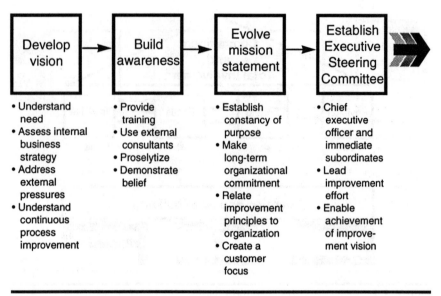

Figure 3.3   Envisioning

tinuous process improvement should become the organization's way of life.

Envisioning, as shown in Figure 3.3, is a process that includes developing the organization's overall mission and goals and, within the context of that overall mission, building individual and group awareness of positive transformation and reengineering objectives, philosophy, principles and practices. Every organization should document its mission and establish the constancy of purpose essential to a successful transformation and reengineering effort. Creating a customer focus is a key element of improving the organization's effectiveness. Each individual must demonstrate belief in the organization's mission and ownership of its vision. An Executive Steering Committee (ESC), led by the head of the organization, guides and leads the overall transformation and reengineering effort, which becomes integrated into the organization's way of doing business. The ESC is also instrumental in enabling the achievement of the mission.

Enabling, depicted in Figure 3.4, is the process by which individuals make it possible for the organization to implement transformation and reengineering principles and practices. It includes

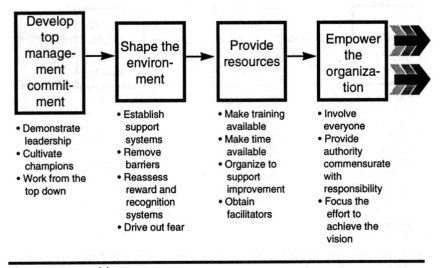

Figure 3.4   Enabling

individual and organizational efforts to create an environment that will support and nurture the improvement effort. Top management must become committed to the implementation and must demonstrate that commitment; highly visible and vocal champions can help publicize that commitment. Everyone must work toward removing the barriers to transformation and establish support, rewards and recognition systems that encourage positive behavior and drive out the inherent fear of change. Training and time resources for the entire work force are essential. Finally, the organization must empower individuals and groups at all levels by providing them the authority necessary to meet their responsibility for process improvement.

Focusing the improvement effort, as shown in Figure 3.5, turns the philosophy and the broad goals into specific objectives and plans for improvement. These goals, objectives and plans are communicated throughout the organization. The transformation effort must ensure that the organization establishes broad, top-level goals and then aligns all of the improvement efforts with those goals. Policy deployment translates broad goals into more specific, relevant goals at each organizational level. Goals at all levels must be realistic, achievable, relevant to both the group and the individual, and

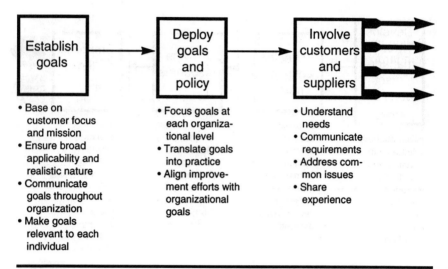

Figure 3.5   Focusing

consistent. The involvement of customers and suppliers ensures that (1) common concerns are addressed, (2) a mutual understanding of everyone's needs exists and (3) information is exchanged in a timely and meaningful manner.

Improving the processes, illustrated in Figure 3.6, is the result of envisioning a new way of doing business, enabling that vision and focusing the effort to achieve specific goals and objectives. The organization's improvement activities include many of the more mechanical processes to define and standardize processes, to assess performance and to improve processes. Performance and progress measurement are critical elements throughout the continuous improvement process. The overriding characteristic of the transformation and improvement process is the establishment of, and adherence to, a structured and disciplined process improvement methodology which allows maximum advantage of individual and collective experiences and energy, as well as institutionalizing that advantage for the good of the organization.

Learning, as shown in Figure 3.7, is one of the fundamental elements supporting the organizational transformation and continuous improvement effort. It comprises training and education. In short, the learning objective should be to provide each individual

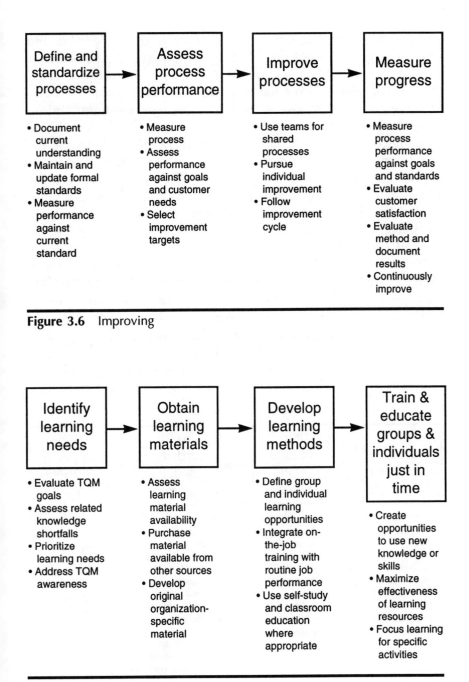

| Define and standardize processes | → | Assess process performance | → | Improve processes | → | Measure progress |

- Document current understanding
- Maintain and update formal standards
- Measure performance against current standard

- Measure process
- Assess performance against goals and customer needs
- Select improvement targets

- Use teams for shared processes
- Pursue individual improvement
- Follow improvement cycle

- Measure process performance against goals and standards
- Evaluate customer satisfaction
- Evaluate method and document results
- Continuously improve

Figure 3.6   Improving

| Identify learning needs | → | Obtain learning materials | → | Develop learning methods | → | Train & educate groups & individuals just in time |

- Evaluate TQM goals
- Assess related knowledge shortfalls
- Prioritize learning needs
- Address TQM awareness

- Assess learning material availability
- Purchase material available from other sources
- Develop original organization-specific material

- Define group and individual learning opportunities
- Integrate on-the-job training with routine job performance
- Use self-study and classroom education where appropriate

- Create opportunities to use new knowledge or skills
- Maximize effectiveness of learning resources
- Focus learning for specific activities

Figure 3.7   Learning

and group exactly the right amount of the correct education and training at just the right time. Doing this requires (a) identification of projected needs from awareness through specific technical skills, (b) determination of how the education and training will be delivered (i.e., in a classroom, on the job or through self-study) and (c) acquisition of the necessary materials and resources. Learning should be planned so that each person and group will be able to use the new knowledge almost immediately after it is acquired. If the learning is not used right away, most people will forget it rather quickly, and a valuable resource will have been wasted. Learning, accomplished in different amounts in different subjects at different times for different people, is necessary through each of the four phases of the LMI CIP Transformation Model.

Team building, depicted in Figure 3.8, is the other fundamental element that will support the transformation and continuous improvement effort. The change and transformation process will gain much of its power and momentum through the formation and activity of teams at all levels in the organization. Teams should be formed according to overall organizational goals, and it is important to ensure that the teams have the necessary training and time resources to work effectively. Team building begins with the estab-

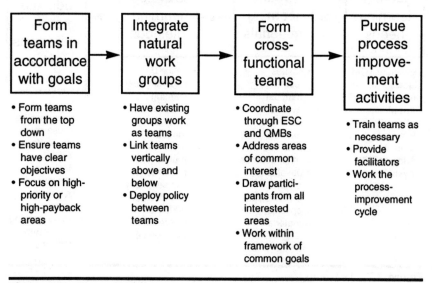

Figure 3.8   Team Building

lishment of the ESC and continues through all levels to the bottom of the organization. In many cases, team building simply means training existing work groups to act as teams; in other situations, common problems and concerns may be addressed through creating cross-functional teams which draw participants from all interested areas. All teams should be linked, horizontally and vertically, and should follow the structured process improvement cycle within the framework of the common organizational goals.

DSMC Q&PMP Transformation Model*

The DSMC Q&PMP Transformation Model, shown in Figure 3.9, is a broad conceptual model with interrelated actions and emphases that describe a general process for transformation from the point when an organization recognizes a need to change to the point at which it becomes a competitive organization of the future. The model depicts an organization as an open system with various feedback loops from the environment, and it highlights the interrelationships between the various components of a quality and productivity management effort.

Organizational System

The "organizational system" box in the middle of the model represents the system that exists; it could be an entire company, a division, a plant, a department or just individual day-to-day activities. The system has upstream systems (internal and external suppliers) which provide inputs in the form of labor, material, capital, energy and data/information. The system takes these inputs and converts them into outputs in the form of products or services. Downstream systems (internal and external customers) then react to those outputs, creating outcomes, such as customer satisfaction, readiness, profitability, etc.

---

* The source of the DSMC Q&PMP Transformation Model is *Quality and Productivity Management Practices on Defense Programs*, Fort Belvoir, Va.: Defense Systems Management College (1988).

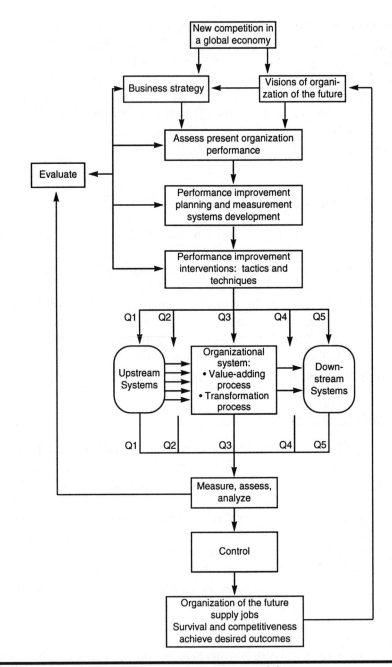

Figure 3.9   DSMC Q&PMP Transformation Model

Incentive and Strategies for Change

At the top of the diagram is the new competition the organization must respond to in order to compete in a global economy. This new competition and global economy influence the business strategy and visions of the organization and the future. The present organizational performance is assessed and the data are used as a foundation for developing plans for performance improvement. Key performance indicators are identified to provide feedback on performance progress.

The following steps are included in an effective strategic planning process: (a) develop a collective strategic awareness among the management team; (b) convert that awareness into specific planning assumptions; (c) create a set of agreed-upon, prioritized, strategic objectives; (d) focus those objectives into a series of action items; (e) determine who will be accountable and responsible for each action item and develop teams to take action; (f) measure, assess and evaluate the effectiveness of improvement actions and (g) continuously support the improvement effort.

Performance Improvement Methodology and Techniques

Out of the performance improvement planning process comes specific performance improvement interventions, tactics and techniques. Note that these interventions happen at five checkpoints: upstream systems, inputs, process, outputs and downstream systems. Quality management efforts must be defined relative to these five checkpoints. In effect, transformation and continuous improvement efforts are commitments to a practice of managing all five quality checkpoints. The management team then develops, through the performance improvement planning process, a balanced attack to improve total system performance, not just system subcomponents.

Measurement and Evaluation

After interventions are made to the system, measure, assess and analyze performance at the five checkpoints to determine whether the expected impact actually occurred. Based on these data, make an evaluation relative to the business strategy, the environment (both internal and external), the vision, the plan and the improvement actions themselves. Note that the process of evaluation is separate from the process of measurement. In addition, measurement supports improvement as its primary objective. The organizational system or unit of analysis being measured must be precisely defined in order to avoid confusion. A number of measurement and evaluation techniques may be used in this regard.

If the organization has an effective, high-performance management process in the areas of planning, measurement and evaluation, control and improvement, it will achieve its vision of the future and its desired outcomes over the long term. An integrated approach to continuous improvement is essential to this achievement.

DSMC/ATI Performance Improvement Model*

The DSMC/ATI Performance Improvement Model, shown in Figure 3.10, is primarily a model for creating an improvement project. It has seven steps and begins with establishing a cultural environment and results in implementing a continuous cycle of improvement projects aimed at improving organizational performance.

Step 1: Establish the Transformation Improvement Process Management and Cultural Environment

The transformation improvement process is a total organizational approach toward continuous improvement of products and services. It requires management to exercise the leadership to establish the conditions for the process to flourish. Management must create a new, more flexible environment and culture which will encourage and accept change. The new culture is developed and operated so that all the people, working together, can use their talents to contribute to the organization's objective of excellence. Management

---

* The source of the DSMC/ATI Performance Improvement Model is *Quality and Productivity Management Practices on Defense Programs,* Fort Belvoir, Va.: Defense Systems Management College (1988).

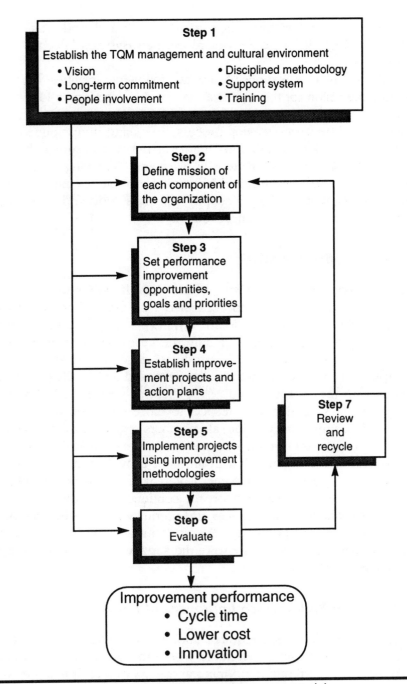

Figure 3.10 DSMC/ATI Performance Improvement Model

must accept the primary responsibility itself and understand the prolonged gestation period before the new systems become alive and productive.

Management is responsible for the following activities: (a) providing the vision for the organization, (b) demonstrating a long-term commitment to implement improvement, (c) actively involving all people in the improvement process, (d) using a disciplined approach to achieve continuous improvement, (e) ensuring that an adequate supporting structure is in place and (f) making all employees aware of the need for, and benefits of, continuous improvement and training them in the philosophy, practices, tools and techniques that support continuous improvement.

Step 2: Define the Mission

The mission of each element of an organization must reflect a perspective such that, when combined with other elements of the organization, it will provide the synergy that produces positive performance improvement. Identify the customer(s), their requirements, the processes and the products; develop measures of the output that reflect customer requirements; and review the preceding steps with the customer and adjust them as necessary. Define the organization's mission with respect to those characteristics.

In developing this mission, all members of the organization must know the purpose of their jobs, their customers(s) and their relation to others in the organization in terms of providing customer satisfaction. Everyone has a customer (internal or external). One objective of the transformation effort for continuous improvement is to provide customers with services and products that consistently meet their needs and expectations. Everyone must know the customers' requirements and must also make the suppliers aware of those and other relevant requirements.

Step 3: Set Performance Improvement Goals

Improved performance requires improvement goals. Both involve change. Steps 1 and 2 determine where the organization wants to go, how it is currently performing and what role each member will play in achieving improved organizational performance. Step 3 sets the goals for performance improvement. These goals must reflect an understanding of the organization's process capabilities so that realistic goals can be set. The goals should first be set at the senior management level. They should reflect strategic choices about the critical processes, the success of which is essential to organizational survival.

Middle and line management set both functional and process improvement goals to achieve the strategic goals set by senior management. This hierarchy of goals establishes an architecture that links improvement efforts across the boundaries of the functional organization. Within functional organizations, performance improvement teams provide cross-functional orientation, and the employees on those teams become involved in process issues. Thus, the entire organization is effectively inter-linked to form an ideal performance improvement culture.

Step 4: Establish Improvement Projects and Action Plans

The initial direction and the initial goals set for the continuous improvement teams flow down from, and are determined by, top management. The steering group performs the following activities: (a) develops the organizational transformation philosophy and vision; (b) focuses on critical processes; (c) resolves organizational and functional barriers; (d) provides resources, training and rewards and (e) establishes criteria for measuring processes and customer requirements.

Step 5: Implement Projects with Performance Tools and Methodologies

Improvement efforts follow a structured improvement methodology. This methodology requires the improvement team to define its customers and processes, develop and establish measures for all process components and assess conformance to customer needs. Analyzing the process will reveal various improvement opportunities, some of which will be more valuable or achievable than others. Opportunities are ranked by priority and improvements effected.

The improvement methodology is a cyclic and infinite process. As one opportunity is pursued and improvements implemented, new opportunities are identified and prioritized. Appropriate performance tools are employed at various points in the process.

Step 6: Evaluate

Measurement is an essential element of the transformation and continuous improvement process. It focuses on the effectiveness of improvement efforts and identifies areas for future improvement efforts. A basic need in all improvement efforts is the ability to measure the value of the improvement in units that are pertinent and meaningful to the specific task. For example, one evaluation of the "before" and "after" levels of customer satisfaction following an improvement effort might include the number of customer complaints. The method of the performance improvement should also be evaluated.

Most organizations have existing measures that may be used with little or no modification. No menu of measurements is applicable to all users. The key is to select measures that can be used by work units to manage and evaluate their products and services so that continuous process improvement may be undertaken.

Step 7: Review and Recycle

The continuous improvement process must be a permanent fixture in the organization. Approaches to positive transformation for continuous improvement that have limited lifetimes will become ineffective if left unattended. Review progress with respect to improvement efforts and modify or rejuvenate existing approaches for the next progression of methods. This constant evolution reinforces the idea that continuous improvement through organizational transformation and reengineering is not a "program" but rather is a new expectation for day-to-day behavior and a way of life for each member of the organization.

Summary

Obtaining commitment to organizational transformation and reengineering efforts means that mechanisms are available to address the apprehension and concerns of people. The transition process can best be handled if a respected person is put in charge of the change effort. In order for an organization to sustain a culture of continuous improvement, it must develop consistent, unified vision, goals and objectives. In addition, it must enable and empower the work force to learn, perform and implement new innovative ideas. An ongoing continuous improvement process for envisioning, enabling, focusing and improving all aspects of work elements must be encouraged.

4

Reengineering
Process
Improvement
Models

*T*hose charged with the responsibility of transforming and reengineering organizations can benefit from process improvement models that have been successfully implemented by other organizations. It is cost effective to benchmark the models that have worked well for other organizations so as not to reinvent the wheel.

Models for reengineering and achieving process improvement in organizations are presented in this chapter. The models represent a collection from state-of-the-art literature; they have been successfully utilized in both public and private organizations. Users will find the step-by-step approach for implementing the models useful when analyzing and improving organizational work processes, systems, structures, procedures and programs.

PMI Leadership Expectation Setting Model*

The PMI Leadership Expectation Setting (L.E.S.) Model, shown in Figure 4.1, revolves around the continuous improvement of quality indicators. The primary expectation of this eight-step model is the continuous improvement of processes and systems within an employee's own function. Leaders and co-workers participate and are expected to provide constructive help and support. L.E.S. is predicated on the belief that individuals in organizations are leaders just as much as top management.

Step 1: Develop a Mission Statement

The first step is to develop a personal mission statement which is consistent with the mission of the entire organization. Make it explicit, but remember that it is to be a guideline for future decision making. As a leader, a personal vision and mission must be understood by other individuals in the organization. A vision understood only by one individual will not move others. A feeling of employee ownership in the organization's future must be cultivated.

Step 2: Identify Key Leadership Functions

The main improvement priority should be to focus process improvement efforts on the highest priority functions. To do this, identify the responsibilities of the job which have the greatest effect on the group's results. From this identification, determine specific opportunities for improvement in the personal leadership processes.

Step 3: Identify Improvement Opportunities

In identifying improvement opportunities, understand what the major job functions and tasks are, focus on the customers' requirements and priorities and, using those requirements, define personal improvement efforts. Finally, identify factors that indicate the level of

---

* The source of the PMI Leadership Expectation Setting Model is *L.E.S. Management* by Louis E. Schultz, Bloomington, Minn.: Process Management Institute (1989).

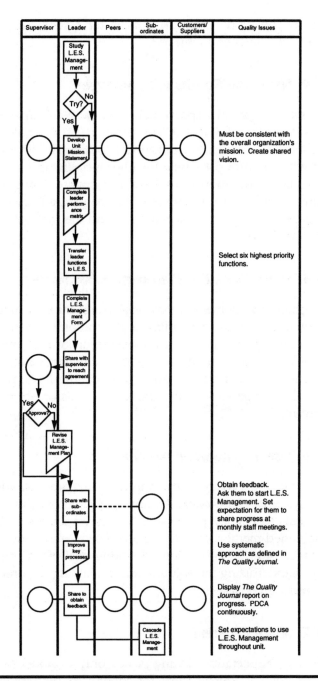

Figure 4.1 PMI Leadership Expectation Setting Model

quality in the work, establish a basis for results and measure the level of improvement achieved.

Step 4: Share the Results with Manager

Discuss the personal improvement plan with a manager to ensure that it is meaningful to the organization as a whole and that it contributes to overall organizational goals. The manager should agree with the proposed plan for improvement and should provide comments on the approach where appropriate. Since the time and resources available for improvement depend to a large extent on the manager's support, it is essential to obtain the manager's agreement before proceeding.

Step 5: Share the L.E.S. Plan with Subordinates

Leadership improvement is meaningless absent the context of those individuals being led. Subordinates are the ultimate purpose of the improvement plan; in effect, they are the major customer. Share the plan with subordinates and ask for their comments and perceptions. Invite them and individual leaders to begin the L.E.S. process themselves. Encourage each individual to share his or her progress with the group.

Step 6: Use a Systematic Approach

To provide consistency to the improvement process, adopt a structured, systematic approach. An approach, such as this one, enables individuals to display progress in a manner understandable by all. A disciplined method of defining a problem, observing it, determining its causes, taking action, checking the effectiveness of that action, standardizing the solution and evaluating the process is a key to providing consistency.

Step 7: Share Progress

Leadership Expectation Setting is not only a model for individual improvement; it is also a basis for continuous performance communication and feedback between employees and supervisors. When sharing progress, do not focus on completing or updating forms; rather, engage in substantive discussion of improvement objectives, obstacles to meeting those objectives and lessons learned along the way. Often, more important lessons are learned in failure than in success; therefore, the performance assessment, both with the supervisor and with the subordinates, should focus on the underlying causes of failure instead of the fact of failure itself.

Step 8: Cascade L.E.S. Management through the Organization

Set the expectation that L.E.S. and individual improvement can be applied at any level in the organization. However, personal leadership and adherence to the process are crucial to its success in the organization. Demonstrate a belief in, and commitment to, the improvement process to help inspire its adoption by others.

Edosomwan Production and Service Improvement Model (PASIM)

The Production and Service Improvement Model (PASIM) is a disciplined process improvement approach which requires continuous process assessment and an organized use of common sense to find easier and better ways of doing work, as well as streamlining the production and service processes to ensure that goods and services are offered at minimum cost. The PASIM concept is shown in Figure 4.2. The PASIM improvement strategy focuses on the following areas:

- Elimination of bottlenecks

- Reduction in production costs, wasted materials, engineering changes, non-value-added operations, the amount of paperwork, chronic overtime, error rate, work repetition, work-in-process inventory, transportation and materials handling, and training time

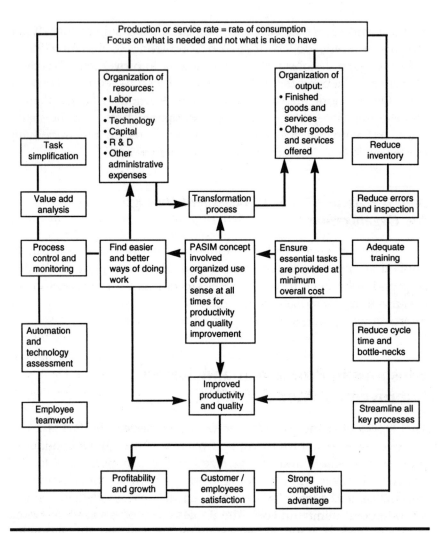

Figure 4.2   PASIM Conceptual Framework

- Improved job safety, employee morale, customer service, and productivity and quality at the source

The PASIM principles are as follows:

Principle 1:   Management and employees must have the positive attitude that productivity and quality improvement can result from the organized use of common sense to address service and production problems. Management support must be shown through example, practice and an organization policy statement.

Principle 2: A total teamwork approach among functional organizations, such as research and development, marketing, personnel, purchasing, manufacturing, information systems, quality, facilities and distribution, maintenance, finance, production control, service centers, engineering and others, must be used to address all problems.

Principle 3: There must be total productivity and quality improvement at the source of production and service. Heavy reliance on inspection and other non-value-adding operations within the work organization must be discouraged. The required basic training must be provided to obtain high-quality goods and services at the source of production or service.

Principle 4: Reduction in the layers of management at all levels must be encouraged. Ownership must be given to those charged with task responsibility, and they must have the control and support needed to resolve daily service and production problems. Too many levels of management causes additional bottlenecks. A level of management must be instituted within the organization only if it will add value to the improvement of the production and service function.

Principle 5: A total impact assessment must be done for all service and process changes, policy changes and implementation of new ideas and techniques. The new process, service, idea, organization or management must be better than the one it replaces, considering all implications.

Principle 6: A total reward system, based on contribution to improving and managing all aspects of a task to obtain acceptable goods and services, must be in place. There must be pay for performance in total technical and people management, and a crisis loop reward system that affects the morale of the employees must be avoided.

Principle 7:   Production and service errors that affect productivity and quality are controllable through common sense and good judgment. The production rate must be equal to the consumption rate.

Principle 8:   Both management and employees must be encouraged to question every task and job in detail. The ability to wonder about the way tasks are performed, to question each step of the task, the power to generalize problem areas and solution strategies and the capacity to apply solutions to fix the obvious problems must be available.

Principle 9:   Both management and employees must be encouraged to eliminate unwanted processes and procedures, to combine work elements to reduce waste and repetition, to have the courage to change the sequence of events if it leads to improvement and to continually simplify work processes.

Principle 10:   Both management and employees should adopt the practice of seeing every job as make-ready, do it and put it away.

PASIM Implementation Steps

The following steps are recommended for implementing PASIM:

Step 1:   Establish a PASIM team which consists of representatives from all functional areas. Each team member must possess, through training, the ability to resolve problems quickly.

Step 2:   Train the team in PASIM principles and concepts, including problem-solving techniques.

Step 3:   Let the PASIM team select a pilot project for improvement. The project selection is done after a thorough analysis of problem areas and prioritization of impact and potential benefits, which would be derived when the new method is implemented.

Step 4:   The PASIM team reviews operational parameters to understand the current situation. This requires six basic steps:

- Orientation to people, processes and procedures

- Information gathering

- Interviewing

- Preparing work distribution charts

- Preparing process flow charts

- Developing task elements lists

This step should focus on detailed review of (1) tasks performed by individuals, work groups, departments and functions; (2) the methods used and (3) the flow of work from one task to another and among the members of the work group. It is also important to identify the volume of work and the frequency, rate and timing involved in the performance of each task.

Orientation: The orientation phase enables the analyst to know all the key processes, procedures and people in the various organizations. The functions for each work group must be clearly understood. Attention must also be paid to inter-functional dependencies. The objective is to obtain all the facts on the task performed by asking ten basic questions:

- What is it?

- What does it do?

- Who does it?

- What does it cost?

- Where is it being done?

- What is it worth?

- When is it done?

- Why is it done?

- Who has the authority to say so?

- Who has ownership for it?

The interviewing process should be structured to obtain an opinion from every level of employee and management. The important point is to ensure that the analyst works with facts instead of opinions. A scheduled time for interviews using a random selection process should also be encouraged.

Work Distribution Charts, Process Charts and Task Elements: Work distribution charts, process flow charts and task element recording formats are organized ways of recording the information gathered from questionnaires, interviews and process procedures. These charts enable analysts to identify the various task activities to be measured and estimate the time spent by each individual or work group on each task. Flow charts show the sequence of steps required to perform a task. Flow charts also provide the basis for understanding the relationship between the task and the processing times. Task elements enable analysts to determine the extent of skill utilization, work specialization, delays, transportation required and other details necessary to pinpoint and understand the process parameters.

Step 5: Conduct operations analysis of selected focus problem(s) in performing an operations analysis of a pilot project. The following information is required:

- Develop a task flow diagram

- Record the processing times for each task

- Record the setup time for each task

- Identify all transportation logistical data

- Simulate the parameters to understand input and output from each sector and leverage area. Simulation also identifies the impact of changes in processing times on other parameters.

The analysis of the information specified above provides the basis for determining bottlenecks, rework loops, capacity limitations and resource constraints. Table 4.1 summarizes the four major questions that need to be answered and the expected results.

Step 6: Analyze current operations to determine the magnitude of problems identified. The feasibility and suitability of all parameters should be determined based on cost, value added, time and impact on quality. Other factors to consider are work safety, job satisfaction and employee morale.

Step 7: Analyze each task for improvement. Six approaches are recommended for operations and service improvement:

- Eliminate and minimize the number of task elements within a given operation.

- Maximize the use of all resources available.

- Combine and rearrange the sequences of processes.

- Substitute and simplify methods of performing a given task.

- Change the sequence for performing a given task.

- Use a new technology or tool to replace the method of performing a given task.

Step 8: Implement the new methods and techniques and evaluate their effectiveness. The teamwork approach should be used for implementation to minimize resistance to change. Evaluation of the new method or technique should be done using the following criteria:

- Cost and savings related to the specific alternative

- Quality improvement related to the specific alternative

- Job satisfaction and morale improvement related to the specific alternative

Table 4.1   PASIM Operations Analysis Questions and Results

Question	*Result*
Who is performing the task?	Reveals hidden organizational problems
	Shows division of responsibility and improvement
	Use of authority and potential improvements
	Communications problems and potential improvements
	Skill base and experience
	Level of authority and decision-making power
	Technical vitality
What task is done?	Redistribution of task
	Elimination of waste and unnecessary task
	Proper utilization of skills
	Consolidation of group activities
	Elimination of wasted efforts
	Balancing of group tasks
	Balancing of input and output rate for each sector
	Correct skill deficiencies
	Task dependencies
	Time utilization
How is the task done?	Simplification of tasks
	Prioritization of task elements
	Elimination of excessive delays and transportation
	Rearrangement of work stations
	Use of faster methods or tools
	Changes in processing time
	Work load balancing
	Performance assessment
Where is the task done?	Scrapping unwanted tools and equipment
	Efficient location of work stations
	Elimination of unwanted transportation
	Improvement in morale of employees
	Reduced design hazards at work stations
	Changes in ergonomic arrangements
	Changes in work design
	Changes in tool design
	Changes in task sequence

- Capability of the user to adapt to the specific alternative

- Implementation requirements of the specific alternative

- Time required to meet all objectives specified in the specific alternative

- Conformation of the specific alternative to established policy and standards

- Specific known exposures related to a specific alternative

- Justification of new alternative tools using the total productivity measurement approach

- Justification of a new alternative using cost–benefit analysis and other financial measures

- Total investment required to implement the new method or technique

Use managerial judgment to assess the degree of tangible and intangible benefits.

Step 9: Follow up on open issues and focus on continuous improvement in all sectors of the production and service environment.

Moen and Nolan Strategy for Process Improvement*

The Moen and Nolan Strategy for Process Improvement, shown in Figure 4.3, is an eleven-step strategy centered on the classic Shewhart or Plan-Do-Check-Act (PDCA) improvement cycle. The eleven steps begin with the selection of a process to improve and result in the implementation of a continuous improvement cycle which operates on the process. The model looks at an organization as a network of linked processes run by internal producers and customers. The

---

* The source of the Moen and Nolan Strategy for Process Improvement is "Process Improvement: A Step-by-Step Approach to Analyzing and Improving a Process" by Ronald D. Moen and Thomas W. Nolan, *Quality Progress* (Sept. 1987).

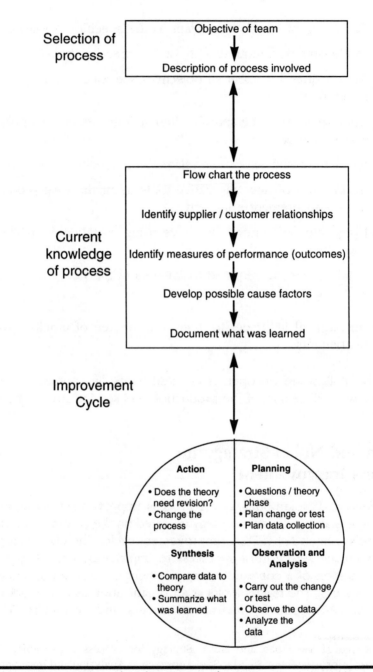

Figure 4.3   Moen and Nolan Strategy for Process Improvement

ultimate output of the network is the product or service provided to an external customer.

Step 1: Determine Team Objective

Identify a process that will have the greatest impact on improving customer satisfaction. The team chosen to work on improving a process should include people working in the process, people in authority to change the process, upstream suppliers, downstream customers and related experts. The team must begin with a clear statement of the objective it hopes to achieve. Each member of the team should view the accomplishment of this objective as important and worthwhile.

Step 2: Describe the Process

Once the team has determined and agreed upon its objective, it should describe and document the process it intends to improve. The documentation should identify all process stages, inputs and outputs. Complete documentation will identify all process suppliers and customers and will attempt to define customer needs and requirements.

Step 3: Flow Chart the Process

One key element of describing a process is creating a flow chart that documents the important stages in the process and identifies relationships between suppliers and customers. The flow chart visually demonstrates the flow of the process over time. Flow charts work best when simple, including only enough detail to give a basic understanding of what is happening. Chapter 5, Tools and Techniques for Organizational Transformation and Process Reengineering, presents more information about flow charts.

Step 4: Identify Supplier/Customer Relationships

Overall performance is improved as producers work in teams with their suppliers (internal and external) to improve internal customer satisfaction and, hence, external customer satisfaction. Suppliers' targets serve as surrogates for customer needs. Each customer becomes the supplier for subsequent needs. Customer and process feedback provides a basis for the improvement action and for measuring subsequent performance.

Step 5: Identify Measures of Performance

Once the team agrees on the flow of the process, it must identify basic measures of performance for the outcome of each stage. These measures are identified as checkpoints on the flow chart. Each measure must be clearly defined as to what specifically is being measured and, more importantly, what that measure means. Identifying performance measures creates windows through which processes can be observed. If those windows do not provide predictable, consistent views of the processes, it will be difficult to make intelligent decisions about how to improve the process.

Step 6: Develop Possible Cause Factors

Measurements provide key indications of process performance problems and their causes. Use a number of tools to keep track of and assess these possible cause factors, which will identify opportunities for improvement.

Step 7: Document What Was Learned

Strict, consistent documentation is essential to maintaining control over the improvement process. Once improvements have been implemented, maintain a history of the entire improvement effort. This history serves to provide lessons which might be applied to other projects and also provides a data trail to analyze the success or failure of the improvement efforts.

Step 8: Plan

Once a project has been selected, the theory phase of the planning step begins. Theory may range from a hunch or "gut feeling" to well-accepted scientific principles at various times throughout the cycle. The next phase is to plan data collection. Data will be used to increase process knowledge and help establish a consensus among team members. The questions to be answered by the data will guide the data collection process.

Step 9: Observe and Analyze

The observation phase begins when the plan for data collection is put in place. The data should be observed as soon as they become available. Any data collection process has many opportunities for error and many opportunities for special causes to occur. Plotting the data chronologically as they are obtained is vital in order to recognize problems.

Once the data are obtained, they are analyzed to help answer the questions posed in the theory phase. In preparing for this analysis, the team should determine the resources needed. Most data from well-planned studies can be analyzed using simple, graphic methods, but there may be occasions when computers are needed. Most teams should quickly learn to use simple tools to collect and display their data. They will usually be able to analyze their own data, but there will be times when help from a statistician or other expert is needed.

Step 10: Synthesize

This phase brings together the results of the data analysis and the existing knowledge of the process. The theory is modified if the data contradict certain beliefs about the process. If the data confirm the existing theory about the process, then the team will be confident that the theory provides sufficient basis for action on the process.

Step 11: Act

Can a change be made in the process or can the process go through its cycle without making a change? If a change is made, will it affect people? What other impact would a change in the process have? These questions and others may be answered by the data collected during the improvement cycle and subsequent analysis. Depending on the answers, process modification may or may not be in order. There is no unique route to problem solving. Agreement on the suitability of improvement action is obtained by repeating the improvement cycle; it is the repeated use of the cycle that is important.

Quality Function Deployment (QFD)

The eight-step process specified in Figure 4.4 is recommended for developing a matrix called the House of Quality. The completed House of Quality should include customer needs and wants as well as design functional characteristics and should depict their relative

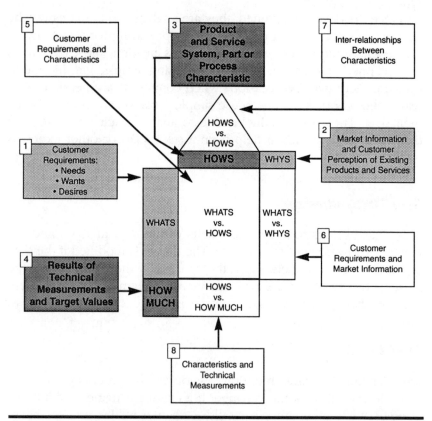

Figure 4.4   Completed House of Quality

importance and the interrelationship between all factors. Each step involved in the construction of the House of Quality is described below.

Step 1: Listen to the voice of the customer and identify his or her needs, wants and desires. Translate the needs, wants and desires into product and service requirements.

Step 2: Obtain marketing information and customer perception of existing and competing products and services.

Step 3: Identify the various product and service system, part or process characteristics.

Step 4: Determine the results of technical measurements and target values.

Step 5: Specify key customer requirements and characteristics.

Step 6: Analyze customer requirements and market information; synthesize data for relevance and accuracy.

Step 7: Develop inter-relationships between characteristics.

Step 8: Fully define product and service characteristics and technical measurement.

Implementing QFD and Resolving Potential Problems

The following steps are suggested for successful implementation of QFD.

Step 1: Incorporate QFD application to the strategic planning process to understand market requirements and strengths and weaknesses of existing products and services and to set the strategic direction for new products and services.

Step 2: Use QFD to define the relative importance of new products and services. Define the product development, manufacturing, marketing and distribution processes with detailed specifications and requirements.

Step 3: Focus on continuous improvement through careful selection of QFD projects and execution. Make sure that the selected QFD project is related to key business needs. In addition, ensure that the selected project satisfies the following criteria: (a) project has direct positive impact on internal and external customers, (b) visible throughout the operational unit, (c) has the endorsement of the total organization, (d) team behind project success, (e) the process involved is stable with its changes, (f) project has start and end points defined, (g) duplication of effort is eliminated and (h) QFD problem statement and expected results clearly understood.

Step 4: Train everyone in QFD concepts and encourage its use in capturing the requirements of the customer at every phase of product or service life cycle.

In order to successfully implement QFD, top management needs to provide leadership and project team effort to use QFD through the entire business process. In addition, top management should support the implementation of QFD by appropriately setting business priorities; training middle and first-line managers in QFD; insisting that QFD designs be based on facts and not opinions; communicating the benefits, pitfalls and the procedures of QFD and encouraging the use of QFD in all business units.

When implementing QFD, one problem that can occur is that it is very easy to mix up requirements. For example, marketing requirements mix with engineering requirements. Requirements definitions must be analyzed carefully and kept separate. Second, QFD may not work if it is applied too late in the development cycle of a product. QFD must be used very early in the product design and development cycle. Third, the QFD requirement matrices can get too large. In order to have a simplified QFD matrix, focus must be placed on the critical attributes. The customer can easily become confused if the questions used to understand the requirements are not stated clearly. Avoid asking customers questions they do not have the technical expertise to understand. It is important to note that QFD can become a meaningless tool if not applied properly to specific projects. Avoid using QFD for every requirement of the business process. Adapt QFD to project needs.

The Quality Journal Model*

The Quality Journal is an adaptation of a Japanese discipline for problem solving. It brings consistency to problem solving in all areas of an organization and displays progress, so that anyone can look at the problem-solving activities, understand the progress and offer additional suggestions for improvement. It is a means for documenting individual improvement efforts. The Quality Journal displays a summary of activities which may be shown in more detail by specific statistical tools. It basically encompasses the seven steps briefly described below.

Step 1: Clearly Define the Problem

Factually state the extent of the problem and how it impacts the total system; then, construct an integrated flow chart to graphically display the process. A problem statement documents in detail what is known about the problem. It explains the reason for selecting the problem, the background of the problem and what has been done to date. An integrated flow chart is a means to examine the process to see what can be done to simplify it, such as the removal of complexity, redundancies and unnecessary actions. The problem-solving effort should be planned, and schedules, time and costs should be estimated.

Step 2: Observe the Problem

Examine the problem from several points of view, which might include different times, places, methods and symptoms. Use specific, focused data collection methods to ensure consistent, accurate and useful data. Involve employees in the data collection process.

---

* The source of the Quality Journal Model is *The Quality Journal* by Louis E. Schultz, Bloomington, Minn.: Process Management Institute (1989).

Step 3: Determine Causes

In determining the causes of the problem, first hypothesize possible causes and then test those causes. Identify possible causes according to main categories established for the problem. Causes that seem to contribute the most to the problem should be noted; information gained from the observation step will be helpful in making this determination. Collect new data to test hypotheses and either prove or disprove them.

Step 4: Take Action to Eliminate Main Causes

Again, use data to evaluate several different solutions to the main causes of the problem. Evaluate and remove root causes and not mere symptoms. Ensure that the solution does not have any detrimental side effects. Finally, select the best solution and implement it.

Step 5: Check Your Solution

Determine the effectiveness of the selected solution. Data are once again the key to this determination. Compare the situations before and after implementation. If the results of the action are not what was desired, first determine if the action was implemented as planned. If so, but the results are undesirable, then test a different solution.

Step 6: Standardize Successful Solutions

After the desired results are achieved, standardize the solution. This involves documenting the successful solution in a new process standard and communicating it to everyone involved in the process. Provide training to ensure the standard is correctly implemented, and devise a system to observe compliance with the new standard.

Step 7: Conclusion

Finally, review the problem-solving procedure and identify any lessons learned about the improvement itself. Note what worked well and what did not, so that future efforts will be better.

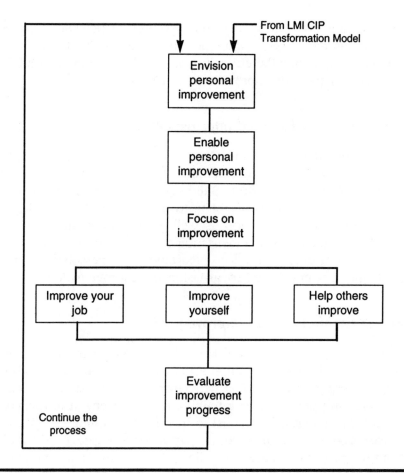

Figure 4.5   LMI CIP Personal Improvement Model

LMI CIP Personal Improvement Model*

The LMI CIP Personal Improvement Model, illustrated in Figure 4.5, follows the basic guidelines of the LMI CIP Transformation Model, but it applies those guidelines to individual improvement efforts. It involves establishing a vision for the personal improvement effort

---

* The source of the LMI CIP Personal Improvement Model is *Introduction to the Continuous Improvement Process: Principles and Pratcices* by Brian E. Mansir and Nicholas R. Schacht, Logistics Management Institute (1989).

and enabling that effort, focusing on personal behavior and the expectations to achieve continuous improvement in performance, and finally evaluating the efforts to improve. The following is a brief discussion of the CIP improvement concepts.

Step 1: Envision Personal Improvement

Before beginning improvement efforts, decide that there is a need for improvement and then determine the general emphasis of the personal improvement effort. Build self-awareness of the need to improve and the individual ability to improve. Assessing relationships within the organization as well as customers and suppliers provides a fundamental understanding of the current status quo. From this assessment, develop expectations for personal behavior and begin creating the personal vision for improvement.

Step 2: Enable Personal Improvement

Make the vision a reality and begin by smoothing the road along which you will travel. This effort starts with educating oneself about improvement goals and about performance improvement concepts, principles and practices. Seek training for improvement in the skills and principles that are essential to the effort. Enabling is a process of learning—learning about using performance improvement tools, about the processes, about the collection and use of data and about the process of learning itself. Also, seek the support of others, not so much from the standpoint of gaining their approval as from the standpoint of cultivating their help in removing barriers to personal effort.

Step 3: Focus on Improvement

Focus on the improvement effort through establishing goals for that effort and then ensure that the improvement activities are aligned with those overall goals. Develop a cohesive improvement strategy to guide the efforts, and ultimately use that strategy to evaluate the success of those efforts. Making improvement a high personal priority and creating time in the schedule for improvement activities are vital to this effort and are a clear demonstration of a personal commitment to improvement.

Step 4: Improve Your Job

Define your job as the collection of the processes owned. Establish control over your job by defining personal processes and understanding how those processes interrelate and relate to others, including customers and suppliers. By removing complexity from the personal processes and pursuing small, incremental improvements, a substantial increase in the effectiveness of personal performance can be achieved.

Step 5: Improve Yourself

Demonstrate leadership in the improvement effort through a commitment to personal improvement. Establish and adhere to a structured, disciplined approach to improvement that clearly defines personal goals and requires steady, consistent improvement performance. Facilitate communication between yourself and others, as well as among others. Remove personal barriers, seek the assistance of others to remove the barriers you do not control and work to eliminate your personal fears of change and improvement. This is best done through education and through communication with others. Depend on the vision to guide the improvements, and use that vision to maintain momentum.

Step 6: Help Others Improve

Through the personal improvement effort, the organization as a whole can improve. An essential part of the personal improvement effort should be to help others improve themselves and the organization. By training and coaching others, by creating more leaders, by working to create teams and eliminate barriers and by encouraging the improvement activities of others, personal examples and enthusiasm will spread throughout the organization. Individuals can make a substantial contribution to the personal improvement efforts of others.

Step 7: Evaluate Personal Improvement Progress

Ascertain the success of the efforts to improve. By measuring personal performance against an established base, by recognizing that the value of improvement lies in the effort to improve beyond the results and by documenting personal improvement efforts so they may be shared with and used by others, you will derive the most from your own efforts. Celebrate personal success and the success of others. Ensure through personal evaluation that the improvement effort itself is rewarding and provides further incentive for continuous improvement effort.

NPRDC Process Improvement Model*

The NPRDC Process Improvement Model, shown in Figure 4.6, is also a PDCA-based model. It begins by stating a goal for improving a process and proceeds through institutionalizing successful process changes in documented process standards.

Step 1: Plan

Select the process to improve and state the goals for that process. Defining those broad goals further, however, requires a description of the process flow by charting the flow itself, documenting the current understanding of how the process functions, defining the customers of the process and understanding customer needs and requirements. Once the process is understood, make the improvement goals more specific; define the actual desired changes in process outcomes. These changes should be realistic, achievable and measurable.

---

* The source of the NPRDC Process Improvement Model is *Defining the Deming Cycle: A Total Quality Management Process Improvement Model* by S. L. Dockstader and A. Houston, San Diego, Calif.: Navy Personnel Research and Development Center (1988).

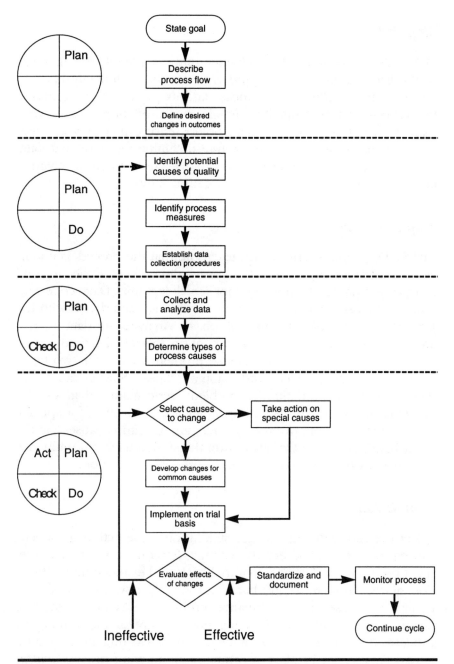

Figure 4.6   NPRDC Process Improvement Model

Step 2: Do

Define the structure for improving the process. Identify the elements of the process, both internal and external, that potentially have an effect on the quality of the process and its products. To verify the theoretical causes of quality, identify measures of process performance. In defining measurement points, ensure that they are specific, repetitive and consistent. Before obtaining measurement data, establish clear, concise data collection procedures to ensure that the data are collected periodically and consistently.

Step 3: Check

Check the process performance to ensure that the process is understood and, more importantly, to improve the process. Collecting and analyzing data is the primary tool for doing this. Data collection must be focused and consistent, performed in accordance with the procedures established in the *Do* phase. Analyze the data aggressively and thoroughly, looking to confirm the initial hypothesis or identify new causes of performance problems. Look for two types of problem causes: special and common causes. Special causes are those sources of variation or problem performance that are not endemic in the system itself, but rather are the result of a specific error in process input or process operation. Common causes, on the other hand, are those that arise from the system itself and influence overall performance in a statistically predictable fashion.

Step 4: Act

Select the causes to be changed, taking one-time action on special causes and developing remedial changes for common causes. Implement both types of actions on a trial basis and evaluate their effects. For ineffective changes, go back and identify new causes of poor quality or causes of performance problems. Document effective changes and build them into the normal way of performing the process; this usually entails modification of existing process standards. Finally, set in place a means of monitoring process performance over the long term, ensuring that the suggested changes continue to have their desired effects and that people are performing the process according to the new standard. The process improvement cycle continues forever...without end.

FPL Quality Improvement Story*

The FPL Quality Improvement Story (QI Story), shown in Figure 4.7, is a seven-part approach which helps to illustrate the steps taken by a team at each stage of an improvement process. It is a standard way of communicating team progress.

The QI Story is utilized because it allows teams to (1) organize, collect and analyze information and provides a way to monitor the team's progress; (2) illustrate and communicate the team's problem-solving process and (3) obtain understanding and input from non-team members.

The QI Story is essentially a structure which allows the team to display the identification, analysis and resolution of a problem or improvement opportunity in a standardized fashion. The QI Story has seven specific steps, each of which contains several logical points which must be dealt with.

QI Story Step	*Logical Points*
1. Reason for Improvement	What indicates there is a need for improvement? Why do we care? What is the effect felt by the customer?
2. Current Situation	Break the theme into its component parts. Stratify the problem many ways to look for the areas of significance. Select the greatest area of significance under the team's control and set a target for improvement.

---

* The source of the FPL Quality Improvement Story is the Florida Power and Light Company, 700 Universe Boulevard, Juno Beach, Florida 33408.

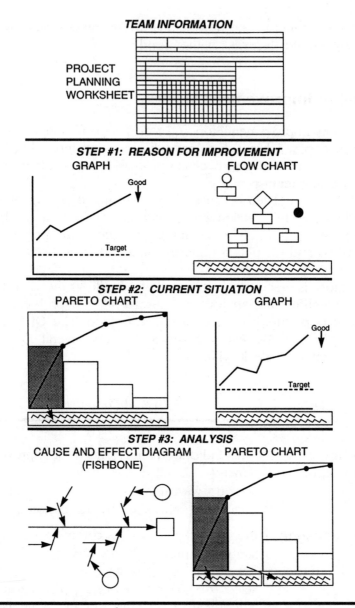

Figure 4.7   FPL Quality Improvement Story

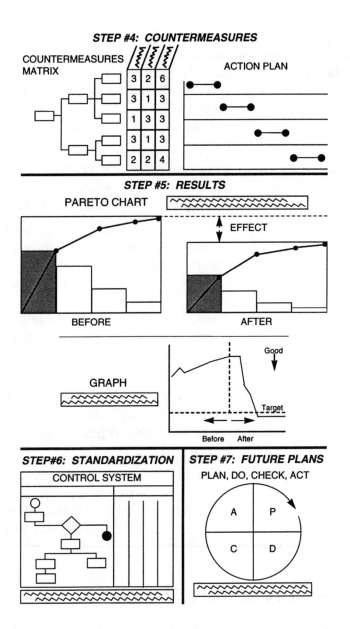

STEP #4: COUNTERMEASURES

COUNTERMEASURES MATRIX

ACTION PLAN

3	2	6
3	1	3
1	3	3
3	1	3
2	2	4

STEP #5: RESULTS

PARETO CHART

EFFECT

BEFORE   AFTER

Good

GRAPH

Target

Before   After

STEP#6: STANDARDIZATION

CONTROL SYSTEM

STEP #7: FUTURE PLANS

PLAN, DO, CHECK, ACT

| A | P |
| C | D |

QI Story Step	*Logical Points*
3. Analysis	Take the most significant area (the problem) and find out why it happens. Verify cause(s) with data and demonstrate their effect on the problem.
4. Countermeasures	Identify possible methods to eliminate the cause(s) of the problem. Select the most feasible and effective, and plan how to implement them.
5. Results	Check to see if the countermeasure(s) were successful in reducing the cause(s) of the problem. Did the overall indicator improve to targeted level? If not, return to the analysis and countermeasures steps to take further corrective action.
6. Standardization	Maintain the gains achieved. Now that there is an improvement, build it into the normal course of business to ensure continued good results.
7. Future Plans	Are there additional benefits to be derived by returning to another portion of this problem? What was learned by solving this problem that will help again next time?

Joiner Associates' Model of Progress*

The Joiner Associates Model of Progress, shown in Figure 4.8, is a six-step process that includes an additional five-stage plan for pro-

---

* The source of the Joiner Associates' Model of Progress is *The Team Handbook*, by Peter R. Scholtes et al., Madison, Wis.: Joiner Associates, Inc. (1988).

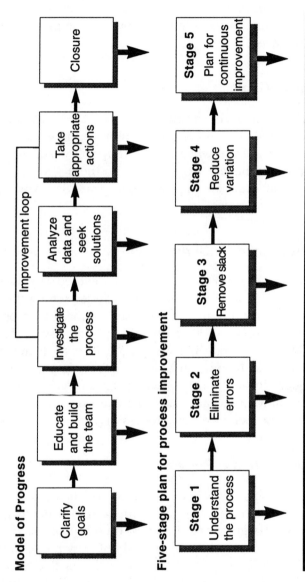

Model of Progress

Clarify goals → Educate and build the team → Investigate the process → Analyze data and seek solutions → Take appropriate actions → Closure

Improvement loop

Five-stage plan for process improvement

Stage 1 Understand the process → Stage 2 Eliminate errors → Stage 3 Remove slack → Stage 4 Reduce variation → Stage 5 Plan for continuous improvement

Figure 4.8 Joiner Associates' Model of Progress

cess improvement. The overall model shows the general progression of events in project teams. It begins by establishing a clear goal based on the organization's mission statement and proceeds through evaluation and management recognition of completed team improvement projects.

Step 1: Clarify Goals

Before the team is even completely constructed, team members begin discussing their mission. They should understand what it means to be on the team, what process they will work on and what kinds of improvements are expected. From these goals and expectations, they will draft an improvement plan that will guide all subsequent team activities.

Step 2: Educate and Build the Team

The first few team meetings are typically devoted largely to team building and education. Team building includes setting the ground rules for team interaction and the logistics for the team's operation. The team should discuss its particular quality issues. The team proceeds with the general discussion of the overall quality philosophy, education and training in specific quality improvement tools and techniques. In addition to developing the technical expertise necessary to improve its processes, the team must take ownership of the process and perceive that process improvement is important.

Step 3: Investigate the Process

After team members have been exposed to quality and scientific principles and have been trained in technical improvement methods, they are ready to begin work in earnest on the process. They begin by studying the process to learn more about how it operates and to identify problems. Process investigation includes documenting the process by using flow charts and diagrams, identifying and communicating with customers and collecting process data. Process data yield clues about root causes of the problems which point to additional data needs.

Step 4: Analyze Data and Seek Solutions

Once the necessary data are collected, the team should analyze them to identify possible causes of problems and then determine which of those possible causes are actually root causes. The five-stage plan for process improvement (see below) helps the team analyze root causes and develop appropriate permanent solutions to the problems.

Step 5: Take Appropriate Action

Once potential solutions have been identified, develop a strategic plan to test the proposed solutions. Implementing the test involves gathering data on the changed process, analyzing the data and redesigning the improvements if necessary. The results of the changes must be continually monitored, not only during the testing period, but permanently. Establish a system with which to monitor improvements as part of the normal way of doing business.

Step 6: Closure

Closure involves presenting the improvement project to management and other interested parties in the organization. It is a means of allowing others to take advantage of the lessons learned and of recognizing team members for their efforts. During closure, evaluate the results of the team's improvement effort and the team's performance during that effort. Finally, complete the documentation of the project.

Plan for Process Improvement

Stage 1: Understand the Process

Before the team can make improvements, each member must thoroughly understand the process. To really know what is right and what is wrong with a process, answer three questions:

- How does the process currently work?

- What is it supposed to accomplish?

- What is the current best-known way to carry out the process?

Investigating these questions is the best way for the team to gather information to set goals and objectives for the rest of the improvement project. Understanding a process is achieved through describing the process, identifying customer needs and concerns and developing a standard process.

Stage 2: Eliminate Errors

Everyone makes mistakes, yet we fail to realize that many mistakes can be prevented by making simple changes to a process. For instance, if people forget to fill in a certain blank on a form or to add the right number of components to a kit, make changes that either highlight the needed step or stop the process until the step is completed. Through actions such as these, it is possible to "error-proof" the process.

Stage 3: Remove Slack

Increasing numbers of organizations are realizing that traditional practices of keeping huge inventories and doing work in large batches are more harmful than helpful. These now-standard practices mask problems instead of solving them. In addition, processes tend to grow over the years; many steps lose whatever value they once had.

To get out of this trap, move toward "just-in-time" flow and examine each step to see if it is necessary and if it adds value to the product or service. The result of this critical examination is an often dramatically reduced time requirement for the completion of a process. The resulting improvements usually increase quality as well.

Stage 4: Reduce Variation

The sources of variation come from both common and special causes; the key is to tell them apart. Common causes typically come from numerous, ever-present sources of slight variation. Special causes, in contrast, are not always present and usually create greater fluctuations in the process. Eliminating common causes requires fundamental changes in how a process is performed; special causes can often be taken care of through relatively simple changes. First focus on reducing sources of variation in the measurement processes and bringing those processes under control, and then focus on performing the same sequence on the targeted processes.

Stage 5: Plan for Continuous Improvement

By this stage, the most obvious sources of the problem will have been eliminated from the process. Now the team must look for ways to make improvement a constant, never-ending part of the process and people's jobs. Ongoing training and education in areas related to the process and instruction in the skills associated with statistical tools are critical. Before bringing the project to a close, discuss ways to keep the improvement philosophy alive. Keep records about the process and procedures up to date; make sure they are used.

LMI CIP Process Improvement Model*

The LMI CIP Process Improvement Model, shown in Figure 4.9, incorporates the PDCA approach but also addresses the need to standardize processes and maintain comprehensive, up-to-date process standards. It begins with the activities needed to create an environment conducive to continuous process improvement, followed by selecting and improving a process, and finally assesses the level of performance improvement; the model then cycles around to focus on another process improvement effort.

Step 1: Set the Stage for Process Improvement

At the organizational level, setting the stage for process improvement involves everything the organization does to become aware of the need for improvement and to establish a commitment to con-

---

* The source of the LMI CIP Process Improvement Model is *Introduction to the Continuous Improvement Process: Principles and Practices* by Brian E. Mansir and Nicholas R. Schacht, Logistics Management Institute (1989).

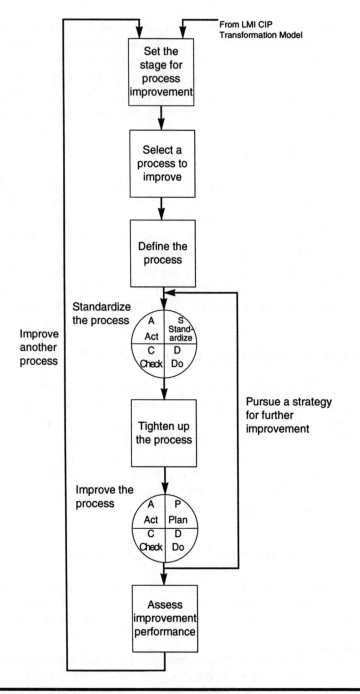

Figure 4.9 LMI CIP Process Improvement Model

tinuous improvement. It includes basic education and training, goal setting, barrier reduction and leadership. Setting the stage means the organization must create an environment in which process improvement activities are encouraged and nourished. The organization must have a clear vision of what it wants to accomplish and where it wants to go, and it must lay in place support systems to help the improvement effort.

At the team and individual levels, setting the stage involves selecting and educating the team or individuals as well as training them in the specific concepts, tools and techniques they will require for the contemplated improvement effort. Determine how the team will function in the overall organizational environment, and ensure that all individuals have enlisted themselves in the accomplishment of their perceived mission.

Step 2: Select a Process to Improve

From all the potential candidates and in conjunction with organizational and team objectives, identify one process on which the team will focus for each pass through the improvement cycle. Selecting the improvement target involves identifying all the potential opportunities, prioritizing them and choosing the process that currently presents the biggest problem or the greatest opportunity for improvement. Once selected, the team must identify the major problems and isolate their root causes. From this background work, the team may create a plan for improvement that builds on the team's objectives. Identifying measurement points is also necessary before beginning the process improvement effort.

Step 3: Define the Process

Once a process has been targeted for improvement, define that process as clearly and completely as possible. Process definition involves determining the customers (both internal and external) and the suppliers of the process, documenting how the process is currently performed (usually by using a flow chart or diagram) and identifying measures of process performance. Documentation should be formal and consistent among all organizational processes.

Step 4: Standardize the Process

By standardizing a process, the best and most current method of performance is institutionalized. Create a means of instructing people in their jobs within a consistent performance definition, provide a means of evaluating performance consistently and provide a basis for evaluating the success of the improvement efforts. Accomplish all this by following the Standardize-Do-Check-Act (SDCA) cycle, which initially requires that measurement systems be brought under control, to identify and document the current method of performing the process (which becomes the process standard) and to communicate and promote use of the standard. Train people in the standard, enable its use and enforce that use. Once the standard is in force, measure all process performance against that standard and respond appropriately to deviations from it. Reducing performance variation by assessing the causes of deviation and eliminating them serves to prevent recurrent deviation. The standard should always reflect the best, most current way of performing the process.

Step 5: Tighten Up the Process

Once a process standard has been defined, tighten up the process before actually attempting to improve it. Tightening up is the maintenance work; make the process improvement effort as effective as possible, ensuring that the process meets its stated and perceived requirements. Cleaning and straightening the process work areas, eliminating unnecessary equipment, instituting total productive maintenance and establishing reliable, adequate data collection systems are essential elements of this effort to tighten up the process.

Step 6: Improve the Process

Efforts to improve the process should follow the classic PDCA cycle: plan for improvement, implement solutions, check for improvement and act to institutionalize improvements. Further efforts will involve developing solutions that address stated requirements and conform to the theories on problem causes. The data collection and measurement methodologies must support whatever solution was envisioned. The team must be trained in the techniques necessary to carry out the plan. After improvement, assess the data to determine how well actual performance matches planned improvements. Successful improvements should be institutionalized; less-than-successful efforts require another pass through the improvement cycle.

Step 7: Assess Improvement Performance

After an improvement has been implemented, document improved performance and the successful improvement effort thoroughly. This documentation allows others to benefit from the lessons the team has learned, and the team gains recognition for its efforts. It also provides a road map to replicate successful improvement techniques. Documenting the improved process also requires updating the process definition and flow diagrams and requires that process standards be rewritten to reflect the new standard of performance. Set in place a means of continuously measuring performance levels if this system does not already exist. Recommend follow-up actions or subsequent improvement efforts. Finally, celebrate your effort!

Summary

Organizations can achieve significant gains in productivity and quality through process improvement. The models presented in this chapter address process improvement at the individual, group, work and organizational levels. The improvement process is not a one-time effort; it is a continuous process that does not have a finish line.

5

Tools and Techniques for Organizational Transformation and Process Reengineering

N o organization or individual can effectively address problems without using the appropriate tools and techniques. The knowledge, experience, tools and techniques of an individual provide the foundation for problem solving and performance improvement.

This chapter presents some of the common tools used in process reengineering. Several factors must be considered before using a particular tool or technique. These factors include problem definition, problem magnitude, problem complexity, problem prioritization, decision analysis, information and data gathering, analyzing alternative solutions to problems, evaluating costs and benefits of projects, planning, measurement, evaluation and improvement of work processes, and addressing behavioral, technical, cultural and management issues. Most tools and techniques are oriented to a certain type of activity, and all tools have a specific purpose, strengths and weaknesses.

Process Analysis Technique (PAT)

This is a systematic approach to defining all tasks required to execute a process (Figure 5.1 and Table 5.1). The following steps are recommended for performing the PAT:

Step 1: Select a particular process based on the degree of performance problem.

Step 2: List the value-added and non-value-added task within the process.

Step 3: Record process times for all activities.

Step 4: Based on the information obtained in Steps 2 and 3, determine which task or activity is required to produce the final output.

Step 5: Seek alternative approaches or methods to perform existing task at reduced cost and improved quality levels.

Step 6: Eliminate the waste and non-value-added tasks, and implement improved value-added methods in the process.

Step 7: Implement the right controls for process monitoring and follow-up on continuous improvement actions.

Flow Charting and Process Analysis Technique

The flow chart shown in Figure 5.2 provides an example of how the various factors and steps can be interrelated in an assembly process. It provides the basis for understanding the standard process procedures and the relationship between the people and the work to be done. When constructed accurately and analyzed properly, a flow chart can help to understand and identify process bottlenecks, such as delays, excessive transportation, waiting time and queuing time. It also is used to identify key customers, suppliers and process owners by operational work-unit performance level, quality level

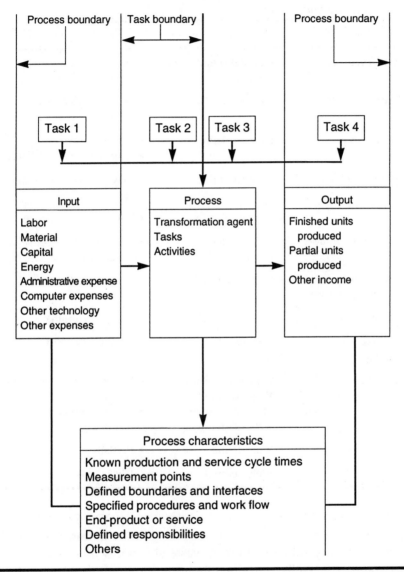

Figure 5.1 The Components and Characteristics of a Typical Process

Table 5.1 Elements of Process Management in
Manufacturing and Service Environments

Element	Manufacturing Sector	Service Sector
• Measurement	• Defined quantitative measures	• Partially quantitative and highly qualitative
• Inputs	• Highly tangible	• Both tangible and intangible
• Ownership	• Easy to define	• Difficult to define • High level of ambiguity
• Boundaries	• Easy to define	• May be unclear in some situations
• Control	• Well established	• Partially in existence
• Corrective Action	• Has defined procedure	• Defined and reactive procedure
• Productivity and Quality Measures	• Highly defined	• Partially defined
• Value Systems	• Mostly tangible	• Mostly intangible
• Process Cycle Times	• Well defined	• Partially defined
• Process Characteristics	• Repetitive tasks • Very common product-oriented technologies	• Non-repetitive tasks very common • Service-oriented technologies
• Customer Base	• Well defined	• Partially defined
• Supplier Base	• Well defined	• Well defined

and productivity at each process point. Flow charting can also be used to identify sources of error, waste and non-value-added operations and to identify introduction steps for new products and services. The following steps are recommended for constructing a flow chart.

Step 1: Understand the process and the relationship between all process parameters (manpower, machines, materials, methods, procedures, technology, systems and policies).

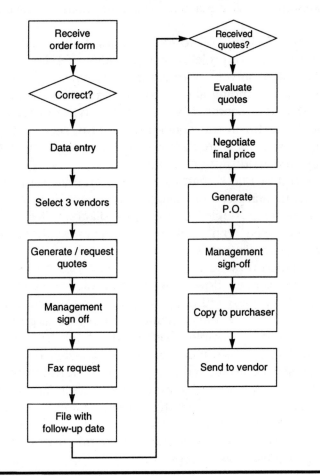

Purchasing Department Process Flow

Figure 5.2 Sample Flow Chart

Step 2: Understand the flow chart process symbols (process, transportation, delays and decision points).

Step 3: Construct the flow chart starting with the first activity or event. Connect all activities or processes using arrows in a chronological order.

Step 4: Identify the key problems by reviewing every step and element specified.

Step 5: Develop a solutions strategy for problems, identifying and implementing corrective actions for continuous improvement.

Flow charts allow you to examine and understand the relationships in a process or project. They provide a step-by-step schematic or picture that serves to create a common language, ensure a common understanding about sequence and focus collective attention on shared concerns. Several different types of flow charts are particularly useful in the continuous improvement process. Three frequently used charts are the top-down flow chart, the detailed flow chart and the work flow diagram.

The top-down flow chart (Figure 5.3) presents only the major or most fundamental steps in a process or project. It helps to visualize the process in a single, simple flow diagram. Key value-added actions associated with each major activity are listed below their respective flow diagram steps.

The detailed flow chart (Figure 5.4) provides very specific information about process flow. At its most detailed level, every decision point, feedback loop and process step is represented. Detailed flow charts should be used only when the level of detail provided by the top-down or other simpler flow charts is insufficient to support understanding, analysis and improvement activity. The detailed flow chart may also be useful and appropriate for critical processes where precisely following a specific procedure is essential. The work flow diagram (Figure 5.5) is a graphic representation or picture of how work actually flows through a physical space or facility. It is very useful for analyzing flow processes, illustrating flow inefficiency and planning process flow improvement.

Work Flow Analysis (WFA)

WFA is a structured system which improves work processes by eliminating unnecessary tasks and streamlining the work flow. WFA identifies and eliminates unnecessary process steps by analyzing functions, activities and tasks. It uses cross-functional teams and is implemented in seven steps (Figure 5.6).

Step 1: Define the process in terms of purposes, objectives and start and end points.

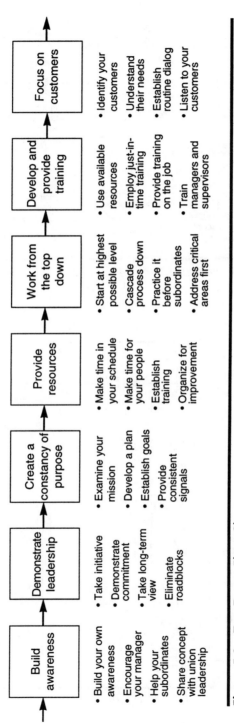

Figure 5.3 Top-Down Flow Chart

Build awareness	Demonstrate leadership	Create a constancy of purpose	Provide resources	Work from the top down	Develop and provide training	Focus on customers
• Build your own awareness	• Take initiative	• Examine your mission	• Make time in your schedule	• Start at highest possible level	• Use available resources	• Identify your customers
• Encourage your manager	• Demonstrate commitment	• Develop a plan	• Make time for your people	• Cascade process down	• Employ just-in-time training	• Understand their needs
• Help your subordinates	• Take long-term view	• Establish goals	• Establish training	• Practice it before subordinates	• Provide training on the job	• Establish routine dialog
• Share concept with union leadership	• Eliminate roadblocks	• Provide consistent signals	• Organize for improvement	• Address critical areas first	• Train managers and supervisors	• Listen to your customers

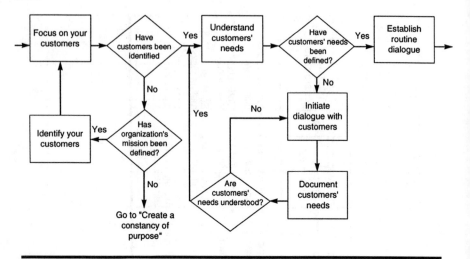

Figure 5.4   Sample Detailed Flow Chart

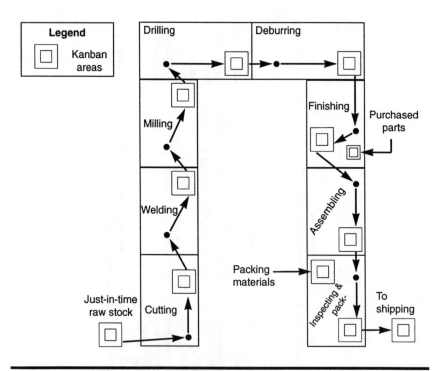

Figure 5.5   Sample Work Flow Diagram

Process / Subprocess Steps	Step 1	Step 2	Step 3	Step 4	Step 5	ETC...
Desired Performance						
Actual Performance						
Cause of Gap						
Recommended Changes						

Figure 5.6 Work Flow Analysis Matrix

Step 2: Identify functions and major responsibilities of the organization, including manpower and planning.

Step 3: Identify activities below functions.

Step 4: Identify tasks or basic steps used to perform each activity and to provide the most specific description of a process.

Step 5: Analyze the process with a cross-functional team.

Step 6: Identify lengthy tasks, choke points, repetitious tasks, etc.

Step 7: Determine and implement an action plan for improvement.

Value Analysis Approach

The value analysis is a systematic approach to examining the functional design of a product or specific part, to develop a more efficient, less costly alternative. The following steps are recommended for implementing the value analysis approach.

Step 1: Select the product or part for analysis and evaluate its importance.

Step 2: Develop a functional definition of the part and describe its purpose and use in the product in question. Specific questions to ask when analyzing parts are:

- Does it contribute to the use of the product?

- Is it cost effective?

- Are all the features required?

- Are there alternative parts that are better?

- Can recycled material be used?

- Can another supplier provide the part for less cost?

- Can the function of the part be combined with something else?

Step 3: Collect data on part performance and cost, and evaluate the contribution of the part to the final product.

Step 4: Develop alternatives. Conduct a brainstorming session to determine the function of the part and to develop ways to overcome any apparent roadblocks.

Step 5: Design or establish specifications for the proposed new component or part.

Step 6: Evaluate the new part through prototype testing, and compare the cost of new and existing parts. Select the best alternative.

Step 7: Implement the preferred part in the product and subassembly, and put it into operation. Follow up to ascertain that the new part is performing the same function at a lower cost and improved quality.

Nominal Group Technique

The nominal group technique (NGT) is a proactive search process that involves a participative group approach to identifying specific

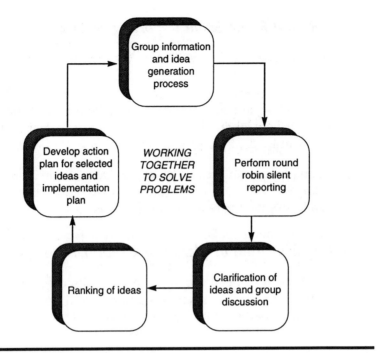

Figure 5.7   Steps for Conducting NGT

problems and issues, or providing ideas and solutions to resolve problems previously identified. It is a methodology for generating ideas, recording those ideas and prioritizing them to move toward consensus decisions. NGT can be especially useful when resolving complex problems, when a group is under time pressures or to avoid potential conflicts associated with discussing and prioritizing sensitive issues. The steps for using NGT are displayed in Figure 5.7 and described below.

Step 1: Idea Generation Process

The group leader presents the purpose of the meeting, which may be to generate ideas for resolving a specific productivity or quality problem. The ground rules for proactive group participation are provided, and time is given for group members to silently record ideas on paper individually without comment.

Step 2: Round Robin Silent Reporting of Ideas

Ideas are collected by the facilitator. Two methods are commonly used:

1. If the ideas are sensitive or if the quantity of participants or ideas is very large, the facilitator collects the ideas and records them individually and anonymously for the group.

2. Each person can present his or her ideas one at a time in turn, with no evaluation or prejudgment by other team members, while the facilitator records the ideas.

Step 3: Clarification of Ideas and Group Discussion

Once all ideas have been recorded, each is discussed for accurate interpretation, to clarify misunderstandings and to combine any ideas that are repetitious in nature.

Step 4: Ranking of Ideas

The ideas are then prioritized by each participant. There are many methods for ranking, the most common being simple voting or weighted voting and Pareto prioritization. The goal of this step is to use ranking techniques to reach a consensus decision by the group on the ideas of interest.

Step 5: Implementation

Once an idea is selected for recommendation or implementation, the group works together to develop an implementation plan and establish an expected results time line.

Fishbone Diagram or Cause-and-Effect Diagram

The fishbone diagram or cause-and-effect diagram, shown in Figure 5.8, helps to relate the elements of a process. It relates possible causes to specific effects. All variation levels are identified by examining all the possible causes. All the possible causes that add to the

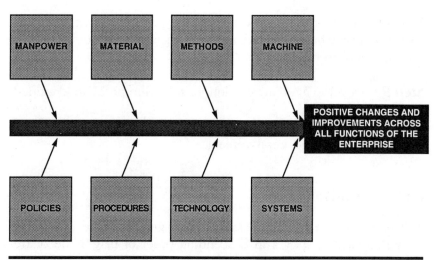

Figure 5.8   Fishbone or Cause-and-Effect Diagram

variation level of the resulting effect are identified using a brainstorming approach. The cause-and-effect diagram provides a method to involve all the people and factors in the service or manufacturing process to see how the various factors come together to make up the total performance.

How to Construct a Cause-and-Effect Diagram

The steps for constructing a fishbone diagram and implementing recommended solution strategies are as follows:

Step 1: Perform a thorough analysis of the production or service operation work unit and define the purpose and problem for using the fishbone diagram.

Step 2: Initiate group meetings involving all parties likely to be affected by the problem.

Step 3: Use the brainstorming or nominal group technique to identify all possible causes of the specific problem and identify potential effect and impact on quality, performance, productivity and total customer satisfaction.

Step 4:   Pinpoint the main causes of the problem. Identify the key contributor (machine, material, methods, technology systems, people, policies or procedures).

Step 5:   Develop alternative solutions to fix the problem identified.

Step 6:   Implement the solution and follow up with continuous corrective action and improvement.

Pareto Analysis

A Pareto diagram can be described as a graphic representation of identified causes, shown in descending order of magnitude or frequency, as depicted in Figure 5.9. The magnitude of concern is usually plotted against the category of concern. The Pareto diagram enables the process improvement analyst to identify the vital key problems, projects or issues on which to concentrate.

How to Construct a Pareto Diagram

The following steps are recommended for constructing a Pareto diagram.

Step 1:   Specify why a Pareto diagram is required, and create a clear definition of the items to be ranked, the criteria to be used and the factor. The motivation for Pareto analysis usually comes from too many complex problems occurring within an operation unit or a specific process. The Pareto diagram is then used to categorize the various problems in their order of magnitude. For example, a computer manufacturer wanted to understand the repair and analysis cost associated with specific types of computer products. The Pareto analysis technique was applied in order to understand the magnitude of this problem. Steps 2 through 4 give specific examples of the application of the Pareto analysis technique.

Step 2:   Perform data collection and record the data by item. In this step, the number of occurrences of each problem and the associated magnitude in weight, cost or time are collected and recorded. For

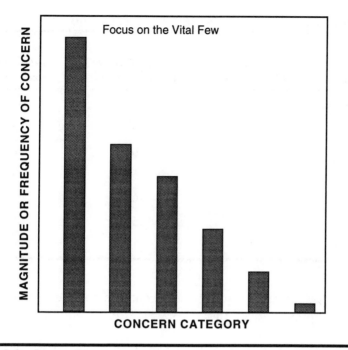

Figure 5.9 Pareto Diagram

the computer products example, the repair and analysis cost for each type of computer is presented in Table 5.2.

Step 3: Calculate percentages for each item and rank them in order. The percentages for each item and the total cumulative percentages for all items are calculated as follows:

$$\text{Item \%} = \frac{\text{each item (weight or value)}}{\text{total items (weight or value)}} \times 100$$

Example for computer type C2:

$$\text{Item \%} = \frac{720}{1515} \times 100\% = 47.5\%$$

Cumulative % = (each item % + previous cumulative %)

Table 5.2   Repair and Analysis Costs for Computer Products

Computer Type	Defect Occurrence (Total Number)	Average Cost per Occurrence ($)	Total Repair Cost per Computer Type ($)
C1	15	20	300
C2	18	40	720
C3	20	10	200
C4	14	5	70
C5	10	10	100
C6	5	25	125
Total	82		1515

The product and cumulative percentages for the computer example are presented in Table 5.3.

Step 4:   Construct graph axes, and plot bars and a cumulative percent line. Based on the values of items obtained in Step 3, the Pareto diagram is constructed. The Pareto diagram for the computer example is presented in Figure 5.10.

Table 5.3   Product and Cumulative Costs for Computer Products

Computer Type	Repair and Analysis Cost ($)	Computer Products (%)	Cumulative (%)
C1	720	47.5	47.5
C2	300	19.8	67.3
C3	200	13.2	80.5
C4	125	8.3	88.8
C5	100	6.6	95.4
C6	70	4.6	100.0
Total		100	

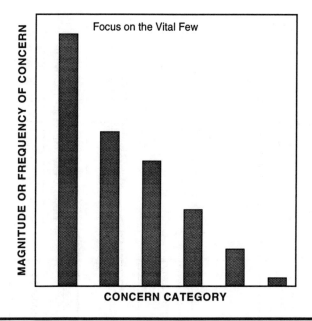

Figure 5.10 Example of Pareto Diagram for Computer Products

Edosomwan Problem-Solving Framework

The Edosomwan General Problem-Solving Procedure, presented in Figure 5.11, is composed of ten steps and is recommended for identifying and solving problems.

Step 1: Problem Identification and Objectives Clarification

In this step, specific problems are identified by examining the current mode of operation. The use of problem identification tools, such as cause-and-effect diagrams, brainstorming techniques and the error removal technique, is suggested. The breadth and scope of the problem should be clearly stated. Data from process flow charts, organization charts, procedures and policies are usually very helpful in specifying the flow of information. Based on the problem(s) identified, specific objectives are formulated. These objectives provide a basis for solution and evaluation.

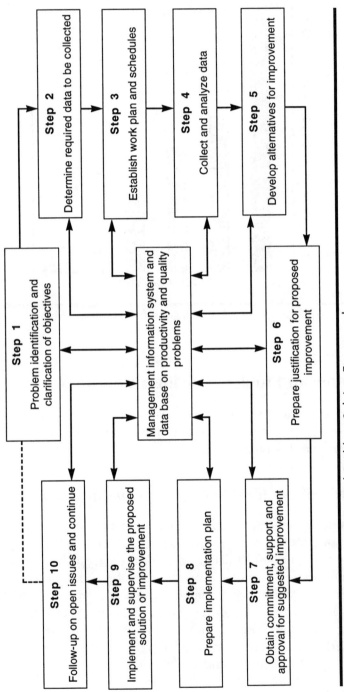

Figure 5.11   Edosomwan's General Problem-Solving Framework

Step 2: Determine Data to be Collected

This is usually accomplished through an understanding of the existing methods for accomplishing tasks. The scope of the problem(s) identified in the first step determines how much data is to be collected.

Step 3: Establish Work Plan and Schedules

It is recommended that the overall problem identified in the first step be broken down into smaller units by tasks and key processes. This will facilitate the analysis of all segments of the problem. Prepare a plan for addressing the problem with a specific detailed work schedule, including distribution of activities to be accomplished by individuals and departments that have responsibility for resolution.

Step 4: Collect and Analyze Data

Ensure that the source of data for the problem being addressed is adequate. Carefully observe the activity under analysis to be sure it is thoroughly understood. Analyzing the data requires attention to data classification; data verification; data synthesis, checking tables, weights and figures; and error verification.

Step 5: Develop Alternatives for Improvement

Re-examine the problem(s) identified in the first step and evaluate the effect of anticipated changes. Collect ideas for solving the problem(s) and improving the situation, and make preliminary plans for improvement.

Step 6: Prepare Justification for the Proposed Improvement

Emphasis is on cost impact assessment, capital investment, rate of return, intangible benefits, and productivity and quality benefits.

Step 7: Obtain Commitment, Support and Approval for Suggested Improvement

Review the proposed improvement with both management and employees affected by the change. Revise improvement alternatives based on their input.

Step 8: Prepare Implementation Plan

The implementation plan should include activities, individual responsibilities and time schedules.

Step 9: Implement the Proposed Solution and Supervise the Installation for Effectiveness

The implementation process should include the training of personnel and resource allocation.

Step 10: Follow-Up

Follow up on open issues and continue to focus on perpetual improvement. Watch closely for erratic changes, perform periodic evaluation of what has been accomplished and continue to seek new improvement opportunities.

Cost of Quality

Cost of Quality is a system which provides managers with cost details often hidden from them (Figures 5.12 and 5.13). It consists of all the costs associated with maintaining acceptable quality plus the costs incurred as a result of a failure to achieve this quality. The cost of not doing things correctly the first time can be considerable. Cost of Quality includes administrative work and is implemented through the following steps.

Step 1: Identify quality costs. These are the cost of non-conformance and the cost of conformance as shown on the diagram.

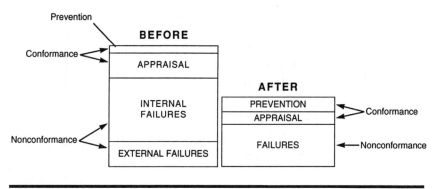

Figure 5.12 Cost of Quality

Step 2: Develop a method for collecting data and reporting on the cost of quality.

Step 3: Identify the most significant costs.

Step 4: Identify the causes of these major costs.

Step 5: Identify solutions to reduce or eliminate causes.

Step 6: Implement solutions.

Team Building

Team building is the development and maintenance of a group of people who can function together to work toward a common goal (Figure 5.14). When a job requires interdependence among the people working on it, it is important to ensure that these people can and will work together smoothly. The six steps of effective team building are as follows.

Step 1: Identify the team.

Step 2: Develop the team: teach group problem solving, openly share data, build norms of shared and collaborative action and teach team members to reinforce one another.

❑ **Prevention:**
- --- Quality engineering
- --- Quality planning
- --- Design verification
- --- Quality training
- --- Quality improvement projects
- --- Statistical process control activities

❑ **Appraisal:**
- --- In-process inspection
- --- Set-up for testing
- --- Administrative costs for quality assurance personnel

❑ **Internal Failure:**
- --- Scrap
- --- Rework
- --- Reinspection of rework
- --- Downtime caused by defects
- --- Investigation of failure or rework

❑ **External Failure:**
- --- Warranty adjustments
- --- Repairs
- --- Customer service
- --- Returned goods
- --- Investigation of defects
- --- Product liability suits

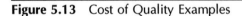

Figure 5.13   Cost of Quality Examples

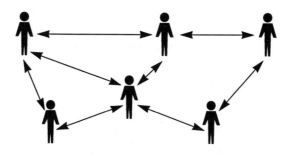

Figure 5.14   Team Building

Step 3:   Identify team goals.

Step 4:   Empower the team members.

Step 5:   Recognize team accomplishments regularly.

Step 6:   Maintain the team.

Force Field Analysis

The force field analysis (FFA) technique is a graphic means of understanding the positive and negative aspects of what prevents us from achieving a goal. FFA assumes that an individual or group knows or is able to recognize the things that will either increase or decrease the likelihood of achieving their intended goal. This tool also focuses on early identification of the stepping stones and roadblocks associated with a problem. This permits a clear focus on a plan of action that will support goal attainment.

FFA should be used to (a) plan the strategy for resolving a problem, (b) collect data and facts for problem analysis, (c) understand the detractors and positive factors influencing a problem, (d) generate and weight alternative solutions to the problem and (e) prioritize and select the most effective solution strategy for implementation. The following steps are recommended for the application of FFA.

Step 1:   Determine the goal and accurately define it.

Step 2:   Using a large notepad (or blackboard), draw a broken vertical line down the right side of the paper, a solid vertical line down the center of the paper and a horizontal line across the top

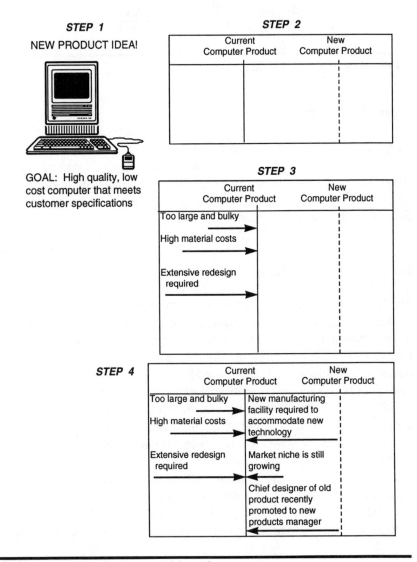

Figure 5.15   Sample Force Field Analysis

of the two vertical lines (see Figure 5.15). The broken line symbolizes the desired state, and the solid line symbolizes the current state. The example in Figure 5.15 describes a "current computer product" as the current state and a "new computer product" as the future or desired state. The notepad serves as the recording instrument.

Step 3: Brainstorm for the items that will help to influence movement from the current situation toward the goal. The example in Figure 5.15 lists three items. Record the items to the left of the "current computer product" line and draw a process arrow from the item to the line. The length of the arrow should be used to approximate the degree of impact that item will have on attaining the goal.

Step 4: Brainstorm for the items that will detract from the process of changing the current situation. Record these items to the right of the "current computer product" line and draw an arrow from the items to the line. The length of the arrow should be used to approximate the degree of impact that item will have on attaining the goal.

Step 5: Discuss and record the strategy, goals and realistic time line for reducing or eliminating each negative force or increasing the position to move toward the desired state.

Summary

In implementing reengineering tools and techniques, the focus should be on eliminating waste, solving problems that affect quality output, maximizing the effective utilization of resources, focusing on the right priorities, customer-driven results, controlling process variation and optimization of organizational performance.

6

Implementing Reengineering Teams and Projects

T eam-directed reengineering efforts encourage more radical and incremental improvement ideas. Using teams to implement reengineering ideas and projects increases ownership, cooperation for implementation tasks, commitment to improvement efforts, motivation, support and buy-in at all levels to achieve results.

Rationale for Reengineering Teams

Implementing a radical or incremental reengineering idea is a difficult task for an individual. When a group of individuals pool their knowledge, skills and talents to work on a reengineering project, implementation of new ideas is faster and more accurate and significant gains in overall performance improvement and results can be achieved. When the reengineering team members represent various functions and departments, there is a depth and breadth of understanding and experience that cannot be matched by a single individual working alone. The team approach allows the transformation and reengineering to be executed in a cooperative manner. It also allows the organization to solve inter-departmental problems that individual departments and functions cannot solve on their own. Utilizing the team approach to reengineering efforts has enormous benefits for customers, suppliers, process owners, employees and the organization as a whole. The benefits of the team approach to problem solving include:

- Gains in performance, quality and productivity

- Continuous encouragement of new ideas and innovation at all levels

- An increase in ownership and accountability of final product

- The creation of a productive work environment which focuses on common vision, mission, objectives, goals and results

In an environment where there is teamwork, people look forward to coming to work each day. People feel good about their work and have a higher self-worth and sense of well-being. Team members look forward to learning something new each day and handling challenges as they arise.

Types of Reengineering Teams

As shown in Figure 6.1, there are four types of reengineering teams. The distinguishing characteristics of the reengineering team are skills, intensity, nature of the reengineering project, experience and inter-dependence. One common characteristic is they are all cross-functional in nature.

Radical Cross-Functional Reengineering Teams

Radical cross-functional reengineering teams are chartered to look at current organizational work processes and totally redesign them

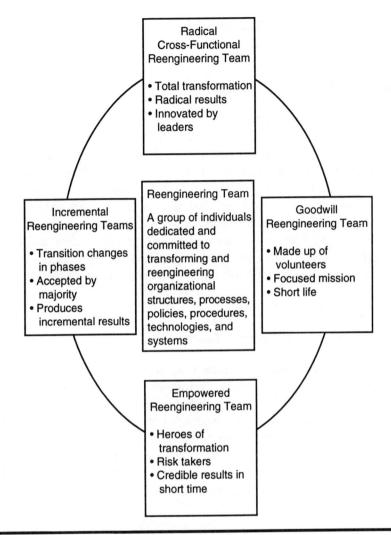

Figure 6.1   Edosomwan Four Types of Reengineering Teams

from beginning to end. These teams utilize transformation and reengineering tools and the problem-solving process to create a radical new organization. They meet regularly to tackle assignments which often require a thorough understanding of total organizational processes and dependencies. The life span of the team can vary from one to six months, depending on the complexity of the reengineering project. The members of these teams are usually appointed by senior management.

Goodwill Reengineering Teams

Goodwill reengineering teams consist of volunteers who have come together for a common purpose. The members are usually dedicated individuals who have strong opinions, convictions and dedication to changing the way the organization does business. These teams are also cross-functional in nature; they meet regularly to work on specific reengineering projects designed by the teams themselves. These teams often run into difficulty with acceptance because they are self-appointed; implementation of their recommendations is usually greeted with skepticism. However, goodwill reengineering teams that can deliver results quickly often gain credibility and the support of senior management.

Empowered Reengineering Teams

Empowered reengineering teams consist of individuals who have achieved tremendous success in transforming and reengineering organizational work processes. These teams are then given the freedom and empowerment to continuously examine how other aspects of the organization are performing and make changes with little or no direct supervision from management. These teams are usually found in high-tech environments where there is a lot of risk-taking, innovation and self-starters who require little or no direction to proceed.

Incremental Reengineering Teams

Incremental reengineering teams are cross-functional reengineering teams that are chartered to analyze an organization's current processes, procedures, policies, systems, technologies and structures and find an incremental solution for the continuous improvement of overall performance. These teams utilize the problem-solving process and tools to create incremental plans and solutions which will transform an organization over a specific time frame. The incremental reengineering team concept is very well accepted in many organizations because the approaches and solutions proposed by these teams often allow people the opportunity to deal with change more comfortably. Incremental reengineering teams are not usually very welcome in an environment where the problems the team is trying to resolve have been around for an extremely long time. When people want immediate solutions and results, they become impatient with incremental reengineering teams. These teams function best in an environment that has adequate strategic planning for continuous improvement.

Guiding Reengineering Teams and Projects

The creation of an Executive Steering Committee (ESC) establishes a high-level senior management group whose focus is to guide the implementation and ongoing direction of the organizational transformation and reengineering efforts, process and projects. The ESC serves as the body that creates strategies, objectives and key performance improvement initiatives and direction. It serves to lend motivation and encouragement to the entire organization as people strive to reengineer their own work processes or support the team effort dedicated to reengineering the organization. The typical ESC consists of representatives of the senior management group, including union leaders. The ESC ensures that the reengineering projects receive maximum cooperation at all levels. The ESC also concentrates on and handles the following tasks and responsibilities:

1. Development and implementation of the transformation and reengineering vision, initiatives, ESC organization and charter, systems, goals, objectives, key results areas and support structure for success.

2. Coordination and support of reengineering and transformance sequence and projects throughout the organization.

3. Identification of key breakthroughs and opportunities for continuous improvement. This responsibility includes identifying problem areas, threats, opportunities and organizational strengths.

4. Education and training in organizational transformation and reengineering. Training should be provided at all levels.

5. Provide the framework and support for effective communication, reward, recognition and ongoing motivation of the work force to support the reengineering efforts and projects.

6. Maintain constancy of purpose to the organizational transformation and reengineering efforts. The ESC constantly reminds individuals and work groups what is important, why it is important, the benefits and the effort required for implementation.

7. Provide resources and hands-on support for reengineering projects. The resources required include manpower, materials, funds, space, technologies, systems and availability of senior management time to review project outcomes and initiatives.

8. Create an ongoing sense of urgency and support for the reengineering efforts. The ESC helps clarify and maximize the role of each level of management, departments and functional areas.

Creation of the ESC is usually the start of an organization's effort to review the initiatives for transformation and reengineering. The formation of the ESC brings together a sense of united leadership to address the core issues and challenges of the organization. It is also recommended that the ESC benchmark other organizations that have undergone transformations and have applied reengineering tools. Learning from the senior management of successful organizations can be very helpful in developing the structure and framework for the ESC and the entire reengineering effort. The ESC teamwork also helps generate support and commitment at the lower levels of the organization.

Creating and Maintaining Reengineering Teams

The reengineering team consists of individuals representing each functional or operational area involved in the organizational transformation and reengineering effort. The reengineering team may be composed of part-time or full-time individuals who represent various interests within the organization. The team works across functional boundaries and meets regularly to identify, analyze and reengineer structures and processes. A successful reengineering team must have the following characteristics: (a) participative approach to goals, objectives and vision setting; (b) open sharing of ideas; (c) open communication among team members and (d) team reward and recognition. The roles and responsibilities of the improvement facilitator, team leader and team members are described in Figure 6.2. It is important to note that the implementation of continuous improvement projects occurs over time. It involves the continuous assessment of progress, accepting the evolution of innovative new approaches, and rescoping and correcting to ensure progress toward objectives. It is recommended that a continuous improvement steering council be formed to monitor and guide project implementation.

The ESC can be involved in determining requirements for tailored orientation and education for all employees, assisting with the development of organized employee action teams and recommending reorganization when appropriate. The ongoing responsibility of the ESC is to assist with the assessment, updating and revision of the continuous improvement strategy as required.

The following guidelines should be considered when establishing reengineering teams. Team members must have the functional or technical expertise required, represent stakeholders interest, have commitment and dedication to the mission of the teams and be available for team meetings, problem solving and resolution. The steps for creating and maintaining a reengineering team are shown in Figure 6.2. It is important for the reengineering team to have clearly defined goals and objectives. The responsibilities of each individual team member should be defined without ambiguity. Each team member should be encouraged to create and focus on the right vision of the organizational transformation and reengineering. Unless the team is empowered for continuous improvement, the formation of a reengineering team must be approved by the ESC. Each team must have a sponsor, who is usually a senior manager.

The discussion regarding team formation should include the objectives and scope of the project, the expected results (including

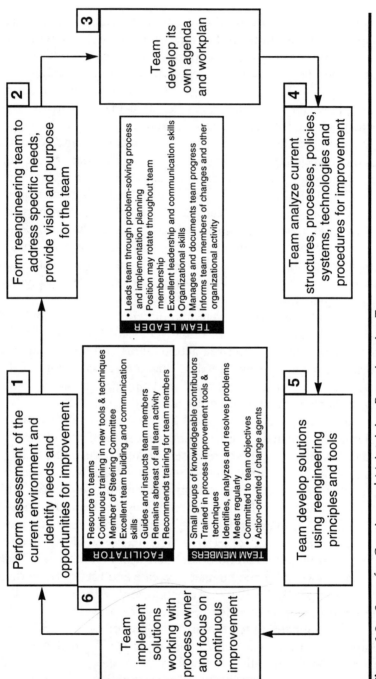

Figure 6.2 Steps for Creating and Maintaining Reengineering Teams

the time frame for completion) and any other relevant details that are available at this early stage. The person who sponsors the request should be prepared to justify the need for the team. One purpose of this discussion is to assure that there is no duplication of team activities (that the project is not already being addressed by another team). Also, this discussion is intended to assure that all areas which need to be involved will be informed and brought into the process at the front end, thus increasing the overall chances for team success. Based on the information that is available, the ESC must determine if there is a need for a team.

It is the responsibility of the sponsor, with input from the stakeholders, to determine team composition. The team should include those who are actually involved with the process to be studied, process customers and suppliers. The ideal team size ranges from five to eight members; however, teams can have as few as four and as many as fifteen members and still be effective. The following should be considered as members are being selected:

- If unions are to be represented on the team, the union leadership should be contacted for recommendations for team members. Those selected by the union must be involved in the process or be suppliers or customers of the selected process.

- Team membership requires technical expertise and experience in process reengineering.

- An individual should not serve as a team member on more than two teams at any one time. When the project is large in scope, restrict individuals to one team.

Once potential team members have been identified, invitation memos should be sent by the sponsor to the members, with copies to their supervisors. This memo should describe the team's project definition and the expectations of the members, including estimated team requirements and team duration.

It is the sponsor's responsibility to meet with the team at its first meeting to discuss the objectives and scope of the project. It is at this meeting that the sponsor must clearly define what is expected of the team and the time frame for completion. The discussion should be very interactive to ensure that all members of the group have a good understanding regarding the team's purpose. After the discussion, the facilitator should assist the team in developing the formal project definition to be used on the team's charter. The team, with the sponsor present, should also discuss the basic steps in the work plan and begin to identify the deliverables, expected resources and projected time frame for completion. Although the project definition, work plan, deliverables and time frame will be preliminary, it is important for the sponsor to be able to give input at this crucial stage.

It is important that the team select a team leader at the first meeting. The team leader should be knowledgeable about the process and have the skills necessary to monitor and keep the team focused on the results.

When assessing the overall benefits of a team, both tangible and intangible results should be considered. Tangible results would include savings incurred due to an increase in productivity, a reduction in process cycle time or a reduction in failure costs. Some of the less tangible results would include the benefits associated with process documentation, the determination of internal and external customer requirements and improved employee morale as a result of increased involvement.

The following suggestions are provided to increase the effectiveness of a reengineering team. At the initial team meeting, the team should discuss the support roles and responsibilities and appoint team members to assume responsibility for each position. The team should select a scribe, leader and facilitator. The team must begin each meeting with a defined agenda and detailed outline of what needs to be accomplished. The agenda should also include topics for discussion, discussion leaders, desired outcomes and the estimated time frame for each topic. At the end of each meeting, the agenda for the next meeting should be drafted. After the meeting, it is the responsibility of the leader, with the assistance of the facilitator, to finalize the agenda and distribute it to team members prior to the next meeting. It is recommended that the agenda for the next meeting be distributed with the minutes of the previous meeting. It is the responsibility of the scribe to take notes during the meeting and document the points in the minutes. The scribe should also keep track of assignments made throughout the meeting and reiterate them at the close. The assignments should be documented in the "Action Items" section of the minutes. At one of the initial meetings, the team needs to determine who should receive the team minutes. The distribution list should include all participants in the project, including the sponsor, the team members and stakeholders.

At one of the initial meetings, the team needs to establish team/ meeting ground rules and determine how to address violations, should any occur. This is important because it eliminates future confusion and problems. Every team member should assemble a Team/Project Binder divided into the following sections:

- Team charter/mission and objectives

- Work plan/project plan/resources required

- Directory of team members

- Meeting schedule (dates/times/locations)

- Progress reports and team meeting minutes

- Current tasks (work in progress/due dates)

- Background information (documents/handouts)

- Deliverables by outcome

- Technical analysis of data and information

- Team presentations

- Recommendations and implementation plan

In order to obtain maximum results from the team, each team meeting should be evaluated to obtain suggestions for improving future meetings. The process for this evaluation should be handled as follows:

- At the end of each team meeting, the facilitator should ask the team members for both positive comments (what went well) and constructive feedback (what could have been done better) on the meeting.

- After the team meeting, the team leader, facilitator and sponsor should conduct a post-meeting analysis to discuss suggestions for improvement of the meetings and team progress.

Implementing Recommendations and Closing Out the Reengineering Team

The recommendations from the reengineering team should be implemented by the process owner, assisted by the sponsor and team members. The process owner should be charged with the responsibility to ensure that the recommendations are implemented and measured to ensure that the expected results are achieved. The reengineering team should be closed out if the team has successfully completed its objectives and has worked with the process owner to implement improvement recommendations. When the team completes its work, appropriate recognition and reward should be provided. The team should be disbanded if it fails to achieve the objectives or redirected to improve performance toward results.

Empowering the Reengineering Team

Empowerment is not a right or privilege. It must be earned. Empowering the reengineering team requires that the following conditions are satisfied:

- **Skills and Ability:** Each member of the team will have the skills and ability to get the job done and implement solutions effectively.

- **Trust and Integrity:** Each member of the team will be trustworthy, able to handle confidential information and will evaluate every situation with the highest professional ethics and moral conduct.

- **Self-Control:**   Each member of the team will have responsibility for and control of the work output, quality defects and problem resolution. Members will inspect their own work and deliver acceptable products and services.

- **Self-Supervision:**   Each member of the team will be able to operate without supervision to produce the desired output. Each team member will not abuse the decision-making latitude provided by the management team.

- **Training and Experience:**   Each member of the team will acquire the appropriate training and experience to achieve self-empowerment for outstanding performance.

Summary

Reengineering process improvement teams play a key role in achieving high-performance results. Before the teams proceed to tackle organizational problems, they must be educated and trained in the use of tools, techniques and approaches for problem solving and decision making. Teams must be encouraged to work together and to share the recognition for accomplishments. The transformation of an organization begins with the transformation of its leadership, individuals and teams. It is not sufficient to create reengineering teams. Every team that is created must have a charter with specific problems and must produce positive results which contribute to overall organizational performance and competitiveness.

7

Success Factors and Addressing Common Implementation Problems

uccessful organizations learn from their own mistakes, benchmark the successful strategies of others, and implement new innovative initiatives to stay ahead of the competition.

Organizational transformation and process reengineering represent radical and incremental changes and a major advancement in the traditional management improvement approaches. This chapter provides key success factors for achieving positive transformation and reengineering organizational work processes. Common implementation problems are discussed, as are strategies for overcoming them.

People Involvement in Organizational Transformation and Reengineering

No organization can successfully achieve a lasting improvement without the involvement of the work force. Total involvement of people is the key to successful implementation of positive improvement in organizations. Total involvement of people involves the engagement of the individual in a personal commitment to assist with all aspects of the required changes. Real involvement requires people to adjust intellectually, physically and psychologically. Involvement requires personal time and effort beyond the simple requirements of the everyday job and involves thinking and caring about the work, the customers, suppliers, the organization and the entire work force. In order to get people involved, they must feel that their decisions count, feel respected and have a sense of worth and dignity. The following suggestions are offered by LMI* to involve the work force in organizational reengineering efforts:

- Establish clear guidelines for manager and employee involvement in improvement activities and provide management training in interpersonal and leadership skills

- Recognize that real involvement in improvement activity is a personal decision, involving some ego risk, and that willingness to take such risk is significantly tied to the organizational environment

- Develop an environment in which the individual is respected and managers listen, help the individual grow and create an atmosphere in which taking risks is encouraged, failure is acceptable and innovation is valued

- Build a strong management ethic under which managers set the example and actively lead the improvement effort

- Develop improvement process sustaining mechanisms to assure that improvement efforts, once started, are not neglected nor allowed to wither

- Make it clear that people are automatically members of process improvement teams because their jobs are part of the process, but allow active involvement in improvement

---

* The source of the work force involvement suggestions is *Introduction to the Continuous Improvement Process: Principles and Practices* by Brian E. Mansir and Nicholas R. Schacht, Logistics Management Institute (1989).

activity to remain a desirable, encouraged, but personal prerogative

LMI also offers the following ideas on how to involve customers, suppliers and unions in the continuous improvement process. Customers are the lifeblood of every organization; understanding their needs and expectations is a precursor to their satisfaction. Involving customers directly or indirectly in the activities and decisions of the organization is an excellent but rarely used vehicle for maintaining awareness of changing customer requirements. Customer involvement can be generated in a variety of ways, including surveys, facility tours, training, joint problem solving, improvement team participation, customized product design effort, suggestion programs, shared service support activity, on-site customer representation and customer recognition programs.

Surveys and market research are among the most common forms of customer involvement. These methods are generally indirect, and customers are not usually aware of the specific initiating organization. Some form of surveying customers and conducting market research is appropriate for most organizations.

Facility tours are another method of involving customers. Tours can be particularly useful if the organization permits the customers to meet with and talk directly to the product designers, assembly line workers or other employees involved in the production and delivery of the product or service. The organization should have a structured means of capturing the ideas and concerns expressed by the customers during the tour. Follow-up question-and-answer sessions at the end of the facility tour may provide a useful forum for collecting customer information.

Customer training offers another excellent opportunity for involving customers and gathering customer information. This training may be specifically product oriented or may be related to general topics of mutual interest to the organization and its customers. Whether the training is conducted at customer premises, at an organizational facility or at a third-party location, a specific mechanism should be available for collecting customer comments, ideas and information. The mechanism might be to involve instruction for data collection or to use post-training questionnaires or follow-up surveys.

Customer problems with products or services offer a particularly important source of customer information and are an important reason for developing positive opportunities for customer involvement. Among the rich sources of customer information are warranty actions, written or oral complaints and customer service hotlines. Each of these mechanisms should have a tie-in to the market evaluation and top management decision-making processes. In particular, these mechanisms lend themselves to statistical analysis of customer problem data as input to improvement analysis activity.

Mutual problem-solving activity is an important means of customer involvement. It is particularly useful when a limited number of customers represent a major share of the market for a product or service. Mutual problem solving may be as simple as informal consultation or assistance or may be as involved as joint design engineering teams. It requires astute management, teamwork and cooperation as compromise may become necessary. The activity should be based on a well-defined problem-solving process, focus on resolving specific issues of concern to customers and have mutually agreed problem identification, objectives and problem-solving strategies defined early in the process.

Mutual problem solving is particularly important when dealing with and involving internal customers. Relationships with internal customers are generally not encumbered by as many sensitive or proprietary issues as is dealing with external customers. Internal customers' processes are frequently closely tied to the producers' processes. Overall process improvement demands communication, teamwork and mutual problem solving. One approach to this issue is through participation in joint process improvement team activity.

Involving customers in process improvement team activity may extend to external as well as internal customers. Customers generally participate by specific invitation rather than by permanent inclusion. A particular customer is included in team activity during those periods when changes are being discussed or actions are being planned and implemented that relate directly to matters of that customer's interest. Different customers may be involved at different periods of process improvement team activity, depending on the goals or projects being pursued.

A customer may be involved in design processes and product configuration decisions, particularly when that customer is the sole or primary customer or when the design is for a customized product. Such involvement is commonplace when the government is the customer, at least with regard to specifications, design review and approval of design changes. The continuously improving organization seeks ways to make such customer involvement a positive win–win teamwork effort rather than a confrontation.

Suggestion programs can be extended to include customers. Customers can be as rich a source of ideas as employees. Mechanisms to ensure rapid and certain response to customer suggestions are essential. Monetary reward for customer suggestions is inappropriate; implemented suggestions usually work to the benefit of the customer in any event. However, some form of recognition, such as a letter of appreciation or a visit from a top-level manager, should be routine.

Exchange programs are another form of customer involvement. Such programs may involve employees spending time working within the customer's activity or customer representatives assigned temporarily to activities within the organization. Such exchanges are beneficial for both parties and could involve either internal or external customers.

Customer service and support activity is another arena for potential customer involvement. Many organizations look to their customers to perform some degree of preventive maintenance or minor repair. This form of involvement can be a rich source of information for the organization. If well managed, it can help create a positive relationship for both parties. Care must be taken to ensure that customer tasks are simple, clean and safe.

Customer reward and recognition may be appropriate in some circumstances. Certainly customers may be invited to organizational award and recognition events. Recognition in the form of a "Customer of the Year" or a "Teamwork Award" might be considered. Such awards are particularly appropriate for internal customers since recognition of internal customers may present fewer difficulties than that of external customers.

Measurement is yet another area of customer involvement, particularly with respect to customer satisfaction. The organization should have a means to measure its performance in meeting customer needs and expectations. Such a system may involve the customer directly or indirectly. The measurement methodology should be well defined and reasonably consistent over time to permit tracking performance or customer changes and trends. Performance measurement involving internal customers should generally be fairly direct and should include a feedback loop that will allow the customer to validate or fine-tune the results of the measurement process.

In order to involve customers in the improvement process, an organization should take the following actions:

- Develop clear guidelines on customer awareness, involve customers in the improvement process and provide training in customer relations to help employees recognize, communicate and develop positive relationships with their customers

- Establish mechanisms to solicit customer ideas, suggestions and concerns, and employ that information in improvement decision making

- Provide customer training that is useful, clear, simple, meets customer needs and involves customers in problem-solving activity, particularly when the problem under consideration directly relates to customer satisfaction

- Include customers, particularly internal customers directly affected, in process improvement team activity and in equipment service activity

- Establish mechanisms to involve customers in performance measurement, tie system performance to customer satisfaction and provide recognition to customers for contributions to the improvement process

Involving the Union

Union participation in continuous improvement projects is crucial to organizations in which a portion of the labor force is unionized. Unions have a history of opposition to major changes in the status quo and in particular to productivity programs. Companies that have attempted to implement changes without first involving union leadership in the earliest stages of the planning process have generally encountered significant opposition.

Union leadership does not wish to abrogate its responsibility for maintaining power, control and influence over the decisions that affect union membership. Historically, union power and influence have flowed from its confrontational relationship with management. Many basic labor benefits have been gained through hard-fought battles of the union against strong management opposition. Involving the union will generally require a careful reshaping of labor–management relationships, greater management willingness to share decision making and a clear demonstration that labor and management share common objectives.

Union leaders representing members in the organization should be invited to participate in improvement projects, in the change education process and in top-level process improvement planning from the outset. The union leadership should be represented at the highest levels of improvement goal setting and be included in the decision processes and communications. Experience has shown that union leadership understands and supports improvement efforts when they have been involved early and have had a role in shaping the deployment process with respect to their membership.

Continuous improvement process principles, when fully understood and practiced, are compatible with the needs and expectations of both management and labor. Higher productivity should not equate with fewer jobs. Higher quality and lower costs should produce more job security and better compensation through higher customer satisfaction and increased sales.

Union leadership involvement should extend to idea-generating schemes and suggestion systems. Organized labor has proven to be a rich source of improvement ideas and problem solutions when given the opportunity and when the change process is beneficial to both management and labor. Responsiveness and recognition regarding union suggestions should satisfy the same criteria applied to employee or customer suggestions.

Including organized labor in the improvement process may require the disclosure of internal information not traditionally shared. Internal management problems are often considered "dirty linen" to be kept hidden. The development of trust is the key to open communication. Most organizations will explore improvement opportunities in low-sensitivity areas first and concentrate on fostering trust-building experiences. Initial successful efforts serve to build a team spirit and open doors for greater engagement.

To involve the union in its transformation and reengineering projects, an organization should take the following actions:

- Establish clear guidelines regarding union involvement, encourage union management cooperation in creating a culture for improvement and promote direct participation by union leadership in improvement process activities from the earliest possible moment

- Provide a means by which the unions can submit ideas and suggestions to the improvement process and ensure that union contributions to continuous improvement are appropriately recognized

- Involve union leadership in management training programs, problem-solving activity and on top-level process improvement teams

- Recognize that the union is in fact a customer for some of the organization's processes and ensure that union involvement extends to performance measurement activity, particularly where the union meets the criteria of a customer

Involving Suppliers

Supplier involvement is an essential ingredient in any complete improvement process. Suppliers play a major role in defining the cost and quality of end products and services. Products and services cannot be substantially and continuously improved without the participation of suppliers.

A successful improvement strategy must include a means for continuously improving the goods and services that are purchased by the organization. That improvement can only be realized if the supplier base is actually engaged in the improvement process. For the improving organization, suppliers become an integral part of day-to-day operations. That closer relationship requires new understandings and new supplier management methods.

This new relationship with suppliers, which continuously improving organizations develop, represents one of the most controversial aspects of continuous improvement initiatives. Continuous improvement forces management to challenge traditional thinking about competition and its effect on purchase price. It brings into question laws and regulations that affect purchasing by the government sector and, in particular, legislation designed to discourage close relationships between buyers and suppliers. Such laws and regulations are intended to prevent corruption, promote fairness, create opportunity for new enterprise and ensure that the government pays the lowest price. However, past performance indicates that such objectives are not always attained and better ways may be available.

Private-sector companies that have developed very close relationships with fewer suppliers have found that equally important advantages can be achieved without necessarily sacrificing the broader objectives sought by the government. Such closer relationships between buyer and supplier have permitted continuous cost reduction through cooperative continuous improvement activity, reductions in variability of purchased materials through better process control and fewer processes, and better investment decision making as a result of longer-term planning. Additional benefits include better production continuity, greater flexibility, shorter lead times and just-in-time inventory management as a result of coordinated production planning and improvement. Improved designs from better design collaboration and lower overhead because of reductions in requirements for inspection, inventory, administrative services and defect correction are also claimed as by-products of a more cooperative relationship between buyer and supplier.

Improving organizations involve their suppliers in training in using continuous improvement tools by educating supplier management teams about improvement processes. They may share their training materials with their suppliers, invite supplier personnel to attend in-house training courses or even conduct training courses in the supplier's facility. They help to facilitate improvement process implementation in supplier activities, including training supplier facilitators and providing facilitation services.

Improving organizations provide meaningful feedback to suppliers on the quality of incoming material as determined by an audit, and they provide that information in sufficient detail and in sufficient time to prevent degradation of incoming quality. When a supplier lacks specific analytical capabilities, the organization may perform a technical analysis and assist the supplier in correcting process problems.

Improving organizations seek to establish joint improvement team activities with suppliers. The organization may include supplier representatives on its process improvement teams and may also supply its own representatives to participate on supplier improvement teams. They engage in joint problem-solving activity and, where appropriate, joint planning and goal setting.

The improving organization develops mechanisms for evaluating and rating supplier management capability and performance using process improvement criteria. It provides feedback to its suppliers on the results of performance evaluations. Suppliers are advised that continuous improvement is essential and that ultimately business will be conducted only with companies that are agreeable to and capable of continuous improvement.

The organization involves the supplier in its idea-generating and suggestion systems and establishes appropriate reward and recognition mechanisms for suppliers to acknowledge their contributions to continuous improvement.

To involve its suppliers in its continuous improvement effort, an organization should take the following actions:

- Establish a goal of ultimately buying only from suppliers that are committed to continuous improvement and quality

- Develop a plan for actively engaging the supplier base in planning, training, problem solving and information exchange

- Design a method for rating and tracking supplier performance, and publish clear guidelines about involving suppliers in the improvement process

- Provide training in supplier relations to help employees recognize, communicate and develop positive relationships with their vendors; establish mechanisms to solicit supplier ideas, suggestions and concerns; and employ that information in improvement decision making

- Provide suppliers with training that is useful, clear and simple and that meets their needs; involve suppliers in problem-solving activity, particularly when the problem under consideration directly relates to customer satisfaction

- Include suppliers as appropriate in process improvement team activity and provide recognition for suppliers, especially in appreciation of their contributions to the improvement process

- Establish mechanisms to involve vendors in performance measurement and tie their performance to ultimate customer satisfaction

Top Management Involvement and Participation

Top management's sincere commitment, support and leadership in continuous improvement efforts is a key requirement for success. This active, participative support and commitment should focus on the eight success factors summarized in Figure 7.1. Top management should recognize the current business environment in order to provide clear goals and objectives for guiding the continuous improvement efforts of the enterprise team. Top management must make a physical, mental and philosophical commitment to quality and total customer satisfaction if there is to be a total or comprehensive quality and customer satisfaction initiative. This type of commitment is generally driven by an awareness and acceptance of the need for change. If there is no strong commitment from top management, priorities will shift before real results can be achieved.

Evidence of commitment is shown through ongoing support for quality and customer satisfaction improvement projects. Top management must monitor, encourage and reward those who are actively improving products and processes. Commitment also involves committing resources to correcting common causes in the system that are responsible for poor quality. When longstanding, systemic issues are identified and require resources to be redirected for resolution, management must respond. It is also important that specific goals and objectives be set to direct employee efforts. Employees must understand how their efforts support the company's strategic plan and must be held accountable for making a contribution.

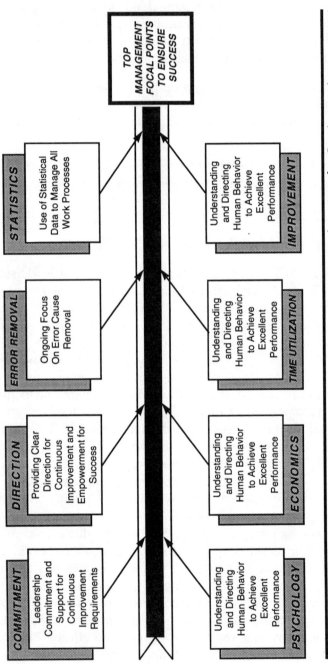

Figure 7.1 Eight Key Focal Points of Top Management Commitment and Support for Continuous Improvement

Top management must also be committed to the utilization of statistics, measurement techniques and data collection to control variability in human, mechanized and automated performance. This represents a philosophical commitment that is key to success. The top management team must continue to grow, develop and change as the culture changes. This growth involves studying, understanding and positively directing human behavior toward excellent performance. Responsibility for managing the economics surrounding the business remains with top management. Costs and prices must be kept at levels that allow affordable products and services to be delivered while the company remains profitable.

Two key elements that must be continually reinforced and demonstrated by top management are a commitment to minimizing delay time, idle time, wasted time, lost time and all other significant aspects of time and to utilizing the methods, procedures and processes required to ensure that the products and services rendered are of exceptional quality.

Finally, top management must be committed to empowering employees. This involves authorizing employees to take corrective action for prevention of errors, defects and delays at every level of the organization. The following steps should be taken to ensure the involvement of top management in continuous improvement efforts.

Step 1: Train top management in continuous improvement tools and techniques, managing change and creating a total quality culture. Involve top managers in continuous improvement training for middle managers, supervisors and other professionals.

Step 2: Make continuous improvement of quality, productivity and total customer satisfaction the chief executive officer's personal mission. Encourage top executive participation in the organization's continuous improvement steering committee to oversee continuous improvement project plans, allocate resources and monitor progress.

Step 3: Require annual continuous improvement plans from line executives and managers. Such plans should depict a blueprint for comprehensive implementation of specific improvement projects. The plan should include specifics on training requirements, customer satisfaction improvement, supplier management, information analysis, process control and data management, employee job satisfaction and human resource issues, benchmarking of competitors and a cost-of-quality estimate for each business unit.

Step 4:   Include discussions about continuous improvement projects in all staff meetings. Top managers should participate in continuous improvement projects, recognition events for quality excellence and the organization's suggestion program for continuous improvement.

Step 5:   Put executives and top managers in touch with outside customers, suppliers and professional organizations. An executive will have a greater appreciation of the requirements, needs and problems of customers and suppliers if one-on-one contact is made. Such contact also provides a unique opportunity for the executive to share and exchange ideas on continuous improvement goals, objectives and specific projects. The outside contact with professional organizations also provides positive exposure for the enterprise and professional validation of new ideas.

The continuous improvement philosophy must be an integral part of top management vision, practice and personal mission.

Blueprint for Continuous Improvement

The development of the organization's improvement strategy should be done with the knowledge gained from the current business environment assessment. A continuous improvement steering council consisting of a group of senior managers and/or their representatives should be used to develop a blueprint for continuous improvement. The continuous improvement steering council serves as the process improvement designer and catalyst for preparing the enterprise for change and providing specific goals and objectives for the improvement. The council also participates in directing the implementation of the continuous improvement projects.

The development of an effective strategy which will direct the entire implementation will require vision and breakthrough thinking on the part of top management. Breakthrough thinking will occur through an effective orientation and education plan for top management on the concepts of continuous improvement. The team should then develop a mission statement and strategic goals with input from employees, customers and suppliers. These goals will be used along with the results of the stage one assessment to develop a phased improvement strategy. The strategy must then be shared with all employees. This stage may take six months to two years. During that time, it is important to keep employees aware of activities and motivated toward change. This stage is probably the most important of all. Developing a clear direction, focus and the collaborative involvement of the top management team are the keys to success.

The blueprint must encompass a long-range strategy for continuous improvement in quality, productivity and total customer satisfaction that covers a minimum of four product cycles or twelve years, whichever is longer. The enterprise management team and non-managers must understand the blueprint to the point that they can develop, at the various operational levels, short-range plans (one to four years) to ensure that their operational activities support the long-range strategy.

These short-range continuous improvement plans should be included in the annual business plan, and each operational unit should be measured throughout the year to determine how well it is meeting these commitments. The measures used at the operational unit level should include, but are not limited to, the following: quality, productivity, customer satisfaction indices, costs, delivery times and schedules, savings per employee, resource improvement ratios and growth in market share. The organization's continuous improvement goals and objectives should be concise, clear and presented in a manner that everyone can understand. Some companies have been able to put their goals and objectives on a small index card that is easy for all employees to understand and remember.

Continuous Improvement Goals and Objectives

It should be recognized that it takes time to involve everyone in the continuous improvement effort. Total quality and productivity improvement cannot be achieved overnight. The implementation of the continuous improvement goals and objectives requires total participation and involvement of the work force. The ability of managers and supervisors to sell the blueprint and the need for continuous improvement is critical. When the management team is involved through hands-on day-to-day participation, employee involvement is easy to obtain. When employees see upper and middle management involvement in, participation in, and dedication to continuous improvement efforts, they will support and participate in them as well.

Selling the continuous improvement goals and objectives to everyone should also focus on why this is needed. Also important considerations are the resources to support improvement projects, assurance of fair reward and recognition for everyone, and the overall benefits for the enterprise. In order to achieve total participation, barriers to communication, both laterally and vertically, must be removed. Total involvement of everyone by providing opportunities to participate in continuous improvement projects is essential to eliminating communication barriers. The focus of the continuous improvement project at the first-line and departmental level should also be to provide a forum for sharing knowledge among team members.

Another method for encouraging employee participation is through a continuous improvement suggestion program. This provides a means for each individual to contribute his or her ideas to the success of the overall goals and objectives established at the enterprise level. Such a suggestion program must also contain measures for success and mechanisms for recognizing individual contributions to improvement.

Education and Training Needs

The assessment of education and training needs should be comprehensive in approach and is a key element of the overall improvement strategy. The assessment should include reviewing the requirements for new hires, formal and on-the-job training approaches and career enhancement opportunities. Interviews would be conducted with employees, including managers, to determine the effectiveness of current training programs and to define training needs associated with continuous improvement process understanding, technical tools, leadership development and participative management. Current training programs should be audited for content and relevance. The recommendations that follow the assessment must include in-house and external training opportunities focused on the continual development of all levels of employees.

Once the education and training needs have been identified, all employees should be trained in the required continuous improvement tools and techniques. While training requirements may vary at the operational and individual levels, it is essential for everyone to make short-term and long-term contributions to the continuous improvement efforts. The training sets the stage for the awareness required for a cultural change. It also provides the required skills and knowledge to address specific problems. The training effort should be extended to suppliers and customers within the framework of business relationships. The key objective of the organization's education and training program is to provide all the requirements for the employees to effectively contribute to the continuous improvement process. The following approaches are recommended for deploying education and training needs.

Strategic: The annual business plan which encompasses the annual quality plan for all levels of the organization should be used for the development of specific types of training for employees to support the continuous improvement initiative.

Universal: The continuous improvement steering committee determines specific, required, ongoing training for all employees, so they will be active participants in the continuous improvement process.

Customized: The management team examines unique or specific needs and then provides customized training utilizing internal or external resources, based on the individual situation.

Individual: Through programs for encouraging active employee participation, all employees review their development needs and, in concert with their managers, individual education and training needs are defined and implemented.

Focus Training Processes and Aids: Most enterprises train the trainers to reduce the overcost of training.

The Human Dimension and Resistance to Change

Process reengineering brings about changes in work habits and work flow patterns; it yields structures that require people at all levels of the organization to readjust their thinking and move away from the current comfort zone to radical new zones that will yield performance improvement. When process reengineering efforts are promoted, the results are often viewed as radical changes; people will either accept or resist the efforts, depending on how the changes will affect them.

One is likely to encounter four types of people when promoting improvement through process reengineering. First, some individuals will do everything in their power to reject the improvement because it creates an extra burden for them. Such people always have several reasons why the new process will not work. They attempt to convince the process improvement agents to remain with the old method of doing things.

The second personality type is those people who do a lot of talking about how process improvement should be done, but take very little action. Third, some individuals will spend their energy wishing someone else or some new system would automatically do the work. Finally, there are those people with a keen interest in improving organizational performance; such people become motivated and accomplish the work, if the necessary support is available. The strategy is (a) to educate everyone about the need for process reengineering and the benefits that it can provide for everyone, (b) provide the vehicle for everyone to participate in the improvement process and (c) reward improved performance.

The Support Structures and Management Commitment

Process reengineering cannot be successful unless the support structures for transformation are fully in place. The strategy is to have short-term and long-term blueprints for achieving transformation and for new process reengineering projects that will optimally satisfy the requirements of the customer and the organization. People involved in process reengineering efforts need to be educated about the transformation process, and requirements and changes need to be made in incremental steps. A strong commitment from and involvement of top management is required if process reengineering projects are to be successful. Commitment from top management should be demonstrated through policies, hands-on participation in improvement projects and rewarding improvement efforts. Top management's willingness to provide the resources required to support process reengineering projects is equally important.

The Complexity of Work Systems and Processes

Process reengineering requires detailed definitions of work elements, tasks, procedures, policies and the sequence of work elements. It also requires the use of internal logic or an external definition of total performance improvement to optimize processes that satisfy customer requirements. Sometimes, this requires substantial changes in the organization. It may require significant resources to evaluate the complexity of work systems. It may also require that reengineering be achieved at a slower pace and in increments.

The Availability of New Technology

Process reengineering requires the use of quality and productivity improvement technologies to achieve a significant breakthrough in complex organizational structures. Often, resources are not available to acquire new technologies which are required to achieve significant improvements. Manual improvement alone is not sufficient. Technologies should be used to automate and reengineer primary, secondary and auxiliary work processes. However, a cost–benefit analysis should be done for each and every tool or technology introduced for process improvement.

Summary

Often, organizational structures and processes are reengineered and redesigned without thorough understanding of the full implications of the impact of the changes and a clear understanding of customer requirements. Everyone involved in the reengineering effort should be trained in process improvement tools and techniques, implementation techniques and managing change. All of the critical voices—unions, suppliers, process owners, management and the work force—must be heard in order to achieve a successful transformation and reengineering of organizations. Top management leadership and involvement must be visible in words, actions and hands-on practical participation in improvement projects. The greatest obstacle to change is human resistance. This obstacle can be minimized by explaining the purpose of the change to people, explaining the benefits that the change will provide for individuals and the organization and involving people in the change definition and implementation process.

APPENDIX

Organizational Transformation and Reengineering Discussion and Tutorial Questions

1. Identify the critical success factors for organizational competitiveness.

2. What are the elements involved in change execution and implementation?

3. Define organizational transformation and reengineering.

4. What elements must be considered in reengineering projects?

5. How would you go about preparing the work force for radical and incremental organizational transformation and reengineering?

6. Define the following terms: process owner, customer, supplier, critical measures, stakeholder and process boundaries.

7. Identify the critical rules for redesigning organizational structures, systems, technologies and processes.

8. What are the six R's of process reengineering?

9. Identify the key cultural changes that are required in organizational transformation.

10. What is the role of global competition in organizational transformation and reengineering efforts?

11. Define the following terms: transformation agent, change agent and process elements.

12. Outline the key performance improvement principles for successful transformation and reengineering efforts.

13. Why is it important to involve stakeholders and unions in the organization's transformation and reengineering efforts?

14. Define the following terms: self-unit customer, supplier and process owner.

15. What are the steps involved in documenting and mapping work processes?

16. Define the Edosomwan tri-level processes.

17. What are the criteria for selecting pilot reengineering projects?

18. Define the key results area and critical success factors for organizational transformation and reengineering.

19. What is the simple rule for evaluating process steps?

20. Why are parallel processes more efficient than linear processes?

21. Identify the primary sources of waste and inefficiency in organizations.

22. Why are people-oriented measures significant in an organization's transformation efforts?

23. How would you go about performing an assessment of an organization's strengths and weaknesses? What instruments would you use and why?

24. Identify the sources of an organization's outdated procedures and policies.

25. What are the key strategies for defining an organization's vision, mission and goals?

26. Why is training of the work force an important element in organizational transformation?

27. Outline the guidelines for handling an organization's transformation process.

28. Identify the Edosomwan Four Stages of Obtaining Commitment to Change.

29. Describe the LMI CIP Transformation Model.

30. Discuss Edosomwan's organizational transformation guidelines. What are the strengths and weaknesses of these guidelines?

31. Why should a respected change agent be put in charge of a reengineering project?

32. What are the significant elements of the LMI CIP Transformation Model?

33. Discuss the key incentives and strategies for achieving successful transformation.

34. Discuss the important features of the Leadership Expectation Model.

35. What are the procedures for developing a personal mission statement?

36. Utilizing the PASIM Model, discuss how operational analysis of work processes should be performed.

37. What are the advantages of using the FPL Quality Improvement Story?

38. How would you go about implementing the Plan-Do-Check-Act (PDCA) improvement cycle in a service environment?

39. Discuss how QFD can be utilized to determine customer requirements.

40. What are the potential problems involved in implementing QFD?

41. How is the Quality Journal Model similar to the Japanese Process Improvement Model?

42. Outline the steps involved in implementing the Joiner Associates' Model of Progress.

43. Why are inspections and transformation steps regarded as non-value-added steps?

44. Outline how the value analysis approach can be used in process reengineering efforts.

45. Outline the steps for conducting a successful nominal group technique.

46. What are the benefits of the brainstorming process in conducting a nominal group technique?

47. Outline how to construct a fishbone diagram. What are the potential uses of this diagram?

48. Why do reengineering and transformation fail?

49. How would you go about prioritizing several innovative ideas for organizational transformation and reengineering?

50. What are the critical elements involved in workflow analysis?

51. Why is the justification of solution alternatives an important step in the problem-solving process?

52. Identify the various types and sources of the Cost of Quality.

53. Present a step-by-step approach for constructing a Pareto diagram.

54. Define reengineering teams.

55. Identify the four types of reengineering teams.

56. What are the roles and responsibilities of the Executive Steering Committee in leading and guiding reengineering teams?

57. What are the potential problems involved in creating too many reengineering projects and teams?

58. What are the roles and responsibilities of team members, facilitators and team leaders?

59. What procedures should be utilized to assess the effectiveness and success of reengineering teams?

60. When should an organization decide to close out a reengineering team?

61. Define the term empowerment. What are the key elements involved in empowering the work force to handle additional responsibilities?

62. Outline the risks involved in empowering the entire work force to handle decision making and problem solving.

63. Why should an organization's leaders involve unions, suppliers, customers and the work force in the transformation and reengineering effort?

64. Outline the procedures for involving external customers in reengineering projects.

65. Why is top management involvement and participation a crucial element in achieving successful organizational transformation and reengineering?

66. Identify the sources of resistance to change and outline strategies for overcoming them.

67. Outline the benefits of using new technologies to achieve organizational transformation and reengineering objectives.

68. Define a process for identifying the training needs for those involved in organizational transformation and reengineering efforts.

69. Define the term change agent.

70. Outline the key steps for developing an organizational blueprint for continuous improvement.

71. Outline the steps for involving senior management in the continuous improvement process.

72. What tools and techniques would you recommend for analyzing customer complaints? For defining customer requirements? For prioritizing customer suggestions?

73. Discuss how the PASIM Model can be used to reduce cycle time and eliminate organizational barriers.

74. Outline key initiatives for rewarding successful reengineering projects.

75. Why is continuous improvement a never-ending journey?

APPENDIX B

Organizational Transformation Change Implementation Checklist

Implementing change successfully requires that the total impact of change on organizational structure, people, business policies, procedures, practices and overall effectiveness be understood. Successful implementation also requires that potential negative or adverse consequences caused by the change be corrected ahead of time. The master of change or change agent seeks answers to the following questions:

1. What has to change in the business environment?

2. What is the magnitude of the change required for success?

3. What are some of the dimensions of the proposed change?

4. How will the change affect the employees, the organization, the management system, the quality of work life and profitability?

5. Who is trained to manage the change?

6. What communication channel will be used for the change?

7. How will everyone identify the mechanisms for involving people in the change process?

8. What mechanisms and avenues are in place to handle grievances, resistance to change and dislocations?

9. How equipped is the management team to deal with the physical, emotional and intellectual aspects of the change?

10. What is the timetable for making the change?

11. What is the anticipated short- and long-term impact of the change on the enterprise?

12. What new skills, knowledge and attitudes are required to make this change?

13. What incentives are available to facilitate the change?

14. What are the potential risks and exposures associated with the change, and what alternatives are in place to counter such risks and exposures?

15. Will the proposed change provide any disincentives for employees? Customers? Other organizations?

16. What reward and benefits are most likely to be achieved by making the change?

17. What are the strengths and weaknesses of the individual/ group undertaking this change?

18. What internal and external obstacles are likely to affect implementation of the change?

19. What are the measures for success?

20. Will the change bring about a win–win situation for everyone?

APPPENDIX C

Process Reengineering Tools and Techniques Matrix

Tools and techniques listed without emphasis on significance/ priority	Area of Emphasis				Applicability Level			
	Problem solving & analysis	Team building	Plan, measure, evaluate	Deployment for results	Task	Process	Individual	Organization
Process Improvement Model								
Process Flow Charting								
Edosomwan PASIT Technique								
Edosomwan Process Reengineering Model								
Edosomwan GM Technique								
Edosomwan Error Mapping Technique								
Nominal Group Technique								
Process Analysis Technique								
Task Simplification and Work Flow Analysis								
Value Analysis Technique								
PMI Leadership Expectation Setting Model								
LMI CIP Process-Improvement Model								
Joiner Associates' Model of Progress								

Tools and techniques listed without emphasis on significance/ priority	Area of Emphasis					Applicability Level			
	Problem solving & analysis	Team building	Plan, measure, evaluate	Deployment for results	Task	Process	Individual	Organization
FPL Quality Improvement Story									
Edosomwan Bench-marking Technique									
Moen & Nolan Strategy Process Improvement									
Cost of Quality									
Edosomwan Waste Analysis Technique									
Force Field Analysis Technique									
Edosomwan Problem Solving Framework									
Pareto Analysis Technique									
Edosomwan Problem Resolution Model									

APPENDIX D

Information Resources

American Center for the Quality of Work Life
37 Tip Top Way
Berkeley Heights, NJ 07922
(201) 464-4609
> Information, training and consulting focused on creating and implementing organizational change and renewal.

American Production and Inventory Control Society
500 W. Annandale Road
Falls Church, VA 22046
(703) 237-8344
> Focus on just-in-time, capacity management, materials requirements, production activity and master planning. Annual conference, seminars/workshops, exhibitions, *Production Inventory Management Journal,* technical publications, professional certification, local chapters.

American Productivity and Quality Center
123 North Post Oak Lane
Houston, TX 77024
(713) 681-4020
> Educational advisory services to organizations in the private and public sectors. Courses, research publications, case studies, *The Letter* newsletter, resource guide, library, consulting.

American Society for Quality Control
310 West Wisconsin Avenue
Milwaukee, WI 53203
(414) 272-8575

The nation's largest professional society dedicated to the advancement of quality. Conferences, educational courses, seminars, *The Quality Review* magazine, journal, book service, professional certification, technical divisions and committees, local chapters.

American Society for Training and Development
1630 Duke Street
Alexandria, VA 22313
(703) 683-8100

Focus on the development and implementation of training programs.

American Supplier Institute
6 Parklane Boulevard, Suite 411
Dearborn, MI 48126
(313) 336-8877

Focus on Taguchi methods, quality function deployment and total quality management. Training, consulting, special conferences, books, videotapes, journal.

Association for Quality and Participation
801-B West 8th Street
Suite 501
Cincinnati, OH 45023
(513) 381-1959

Focus on quality circles, self-managing teams, union–management committees, socio-technical analysis and other aspects of employee involvement. Conferences, library and research service, *Quality and Participation* journal, newsletter, resource guide, local chapters.

Canadian Supplier Institute
Skyline Complex
644 Dixon Road
Rexdale, Ontario M9W 1J4
(416) 235-1777

Focus on Taguchi methods, quality function deployment and total quality management. Training, consulting, books and videotapes, special conferences.

Center for Quality and Productivity
University of Maryland
College Park, MD 20742-7215
(301) 403-4535

Center for Quality and Productivity Improvement
University of Wisconsin-Madison
620 Walnut Street
Madison, WI 53705
(608) 263-2520

Center for the Productive Use of Technology
George Mason University, Metro Campus
3401 North Fairfax, #322
Arlington, VA 22201
(703) 841-2675

Community Quality Coalition
c/o Transformation of America Industry Project
Jackson Community College
2111 Emmons Road
Jackson, MI 49201
(517) 789-1627

Continuous Improvement Company
Efe Quality House
3970 Chain Bridge Road
Fairfax, VA 22030
(703) 359-5970

Georgia Productivity Center
Georgia Institute of Technology
219 O'Keeffe Building
Atlanta, GA 30332
(404) 894-6101

Institute of Industrial Engineers
25 Technology Place
Atlanta, Norcross, GA 30092
(404) 449-0460
Focus on quality and productivity improvement, systems integration and engineering applications for continuous improvement.

Manufacturing Productivity Center
10 West 35th Street
Chicago, IL 60616
(312) 567-4800

Pennsylvania Technical Assistance Program (PENNTAP)
Pennsylvania State University
University Park, PA 16802
(814) 865-0427

QCI International
P.O. Box 1503
Red Bluff, CA 96080
(916) 527-6970
Focus on employee involvement, total quality and statistical process control. Seminars, in-house training, books, videotapes, *Quality Digest* magazine.

Quality and Productivity Management Association
300 Martingale Road
Schaumburg, IL 60173
(708) 619-2909
Network of North American quality and productivity coordinators, operating managers and staff managers. Conferences, workshops, journal, newsletter, resources guide, local chapters.

Work in America Institute
700 White Plains Road
Scarsdale, NY 10583
(914) 472-9600
Research and member advisory services focusing on productivity and quality through employee involvement, labor–management

relations and quality of working life. Member forum and seminars, site visits, policy studies, *Work in America* newsletter, library, research service, technical publications, speakers and consultants bureau.

Consultants

American Productivity & Quality Center
123 North Post Oak Lane
Houston, TX 77024
Contact: Jackie Comola
(713) 681-4020

American Supplier Institute
15041 Commerce Drive, South
Dearborn, MI 48126
Contact: John McHugh
(313) 336-8877

Coopers & Lybrand
1525 Wilson Boulevard, Suite 800
Arlington, VA 22209
Contact: Ian Littman
(703) 875-2102

Ernst & Young
1225 Connecticut Avenue, Northwest
Washington, D.C. 20036
Contact: Renee Jakubiak
(202) 862-6000

Federal Quality Institute
441 F Street Northwest
Washington, D.C. 20001
Contacts: John Franke, Director, (202) 376-3747
        Jeff Manthos, Information Network Office, (202) 376-3753
The Federal Quality Institute (FQI) provides a start-up service in total quality management for top-level federal government man-

agement teams. The FQI serves as a government-wide focal point for information about total quality management through the FQI Information Network, which lends to the "federal" sector materials such as videotapes, books and case studies at no cost.

General Systems Company, Inc.
Berkshire Common, South Street
Pittsfield, MA 01201
Contact: Armand Feigenbaum
(413) 499-2880

Goal/QPC
13 Branch Street
Methuen, MA 01844
Contact: Stan Marsh
(508) 685-3900

IIT Research Institute
Beeches Technical Campus
Route 26 North
Rome, NY 13440-8200
Contact: Steve Flint
(315) 337-0900

Johnson & Johnson Associates, Inc.
3970 Chain Bridge Road
Fairfax, VA 22030
Contact: Wanda Savage-Moore
(703) 359-5969
International consultants in business process reengineering, quality, productivity, customer satisfaction, innovation and technology management and organizational development.

Juran Institute, Inc.
88 Danbury Road
Wilton, CT 06897
(203) 834-1700

Navy Personnel Research and Development Center
Quality Support Center
San Diego, CA 92152-6800
(619) 553-7956
(619) 553-7956 (AV)

Philip Crosby Associates, Inc.
807 West Morse Boulevard
P.O. Box 2369
Winter Park, FL 32790
(305) 645-1733

Technology Research Corporation
5716 Jonathan Mitchell Road
Fairfax Station, VA 22039
Contact: V. Daniel Hunt
(703) 451-8830 and (703) 250-5136

U.S. Army Management Engineering College
Rock Island, IL 61299-7040
(309) 782-0470
(309) 793-0470 (AV)

Awards

Deming Prize Resource Center
Information concerning the Deming Prize application can be obtained by contacting:
Junji Noguchi
Executive Director
Union of Japanese Scientists and Engineers
5-10-11 Sendagaya
Shibuya-Ku
Tokyo 151, Japan
FAX: 9-01-813-225-1813

Malcolm Baldrige National Quality Award
National Institute of Standards and Technology
Route 270 and Quince Orchard Road
Administration Building, Room A537
Gaithersburg, MD 20899
(301) 975-2036
FAX: (301) 948-3716

Publications

American Production and Inventory Control Society
500 West Annandale Road
Falls Church, VA 22046
(703) 237-8344
   Production Inventory Management Journal, quarterly, $110/year.

American Productivity & Quality Center
123 North Post Oak Lane
Houston, TX 77024
(713) 681-4020
   The Letter newsletter, monthly, $125/year.

American Society for Quality Control (ASQC)
310 West Wisconsin Avenue
Milwaukee, WI 53203
(414) 272-8575
   Quality Progress, 12 issues/$40, monthly publication. *Journal of Quality Technology,* 12 issues/$13, monthly publication. *The Quality Review* magazine, quarterly, $36/year.

Association for Quality and Participation
801-B West 8th Street
Cincinnati, OH 45023
(513) 381-1959
   Quality and Participation journal, quarterly, $35/year.

Buraff Publications
2445 M Street Northwest, Suite 275
Washington, D.C. 20037
(202) 452-7889
   Work in America newsletter, monthly, $247/year.

Hitchcock Publishing Co.
191 South Gary Avenue
Carol Stream, IL 60188
(708) 665-1000
Quality magazine, monthly, $65/year.

Manufacturing Productivity Center, IIT Center
10 West 35th Street
Chicago, IL 60616
(312) 567-4808
Manufacturing Competitiveness Frontiers magazine, monthly, $100/year.

PRIDE Publications
P.O. Box 695
Arlington Heights, IL 60004
(708) 398-0430
Commitment-Plus newsletter, monthly, $95/year.

Productivity, Inc.
101 Merritt 7
Norwalk, CT 06851
(203) 846-3777
Productivity newsletter, monthly, $167/year.

QCI International
P.O. Box 882
Red Bluff, CA 96080
Quality Digest magazine, monthly, $75/year.

Society of Manufacturing Engineers
P.O. Box 930
Dearborn, MI 48121
(313) 271-1500
Manufacturing Engineering magazine, monthly, $60/year.

The Quality Observer Corporation
P.O. Box 1111
Fairfax, VA 22030
(703) 691-9496 and (703) 691-9295

The Quality Observer International News Manazine, a leading authority in quality improvement, productivity improvement, benchmarking, reengineering, customer satisfaction, quality awards and programs, and organizational competitiveness. Subscription: $79/year.

APPENDIX E

Reengineering Tools and Packages

- Process Reengineering Assessment Package
- Radical and Incremental Reengineering Tools
- Reengineering Self-Assessment Questionnaires
- Organization Redesign and Reinvention Package
- Staffing Analysis Tool
- Process Analysis Tool
- Job Analysis Tool
- Waste Analysis Tool
- Task Analysis Tool
- Quality Analysis Tool
- Cost Analysis Tool
- Productivity Improvement Tool
- Organization Structure Analysis Tool
- Benefits and Cost Analysis Tool
- Relationship Analysis Tool
- Defect Analysis Tool

Suggested Readings

Adair, Charlene B. and Bruce A. Murray. *Breakthrough Process Redesign*. New York: AMACOM, 1994.

Albrecht, Karl. *At America's Service: How Corporations Can Revolutionize the Way They Treat Their Customers*. Homewood, Ill.: Dow Jones-Irwin, 1988.

Albrecht, Karl and Ron Zemke. *Service America: Doing Business in the New Economy*. Homewood, Ill.: Dow Jones-Irwin, 1985.

Argyris, Chris. *Integrating the Individual and the Organization*. New York: John Wiley & Sons, 1964.

Aubrey II, Charles A. and Patricia K. Felkins. *Teamwork: Involving People in Quality and Productivity Improvement*. Milwaukee, Wis.: Quality Press, 1988.

Balm, Gerald J. *Benchmarking: A Practitioner's Guide for Becoming and Staying Best of Best*. Schaumburg, Ill.: Quality & Productivity Management Association, 1992.

Batten, Joe D. *Tough-Minded Leadership*. New York: AMACOM, 1989.

Bennis, Warren. *The Planning of Change*. New York: Holt, Rinehart, and Winston, 1976.

Bennis, Warren and Burt Nanus. *Leaders: The Strategies for Taking Charge*. New York: Harper & Row, 1985.

183

Camp, Robert C. *Benchmarking: The Search for Industry Best Practices that Lead to Superior Performance*. Milwaukee, Wis.: ASQC Quality Press, 1989.

Collins, Frank C. *Quality: The Ball Is in Your Court*. Milwaukee, Wis.: ASQC Quality Press, 1986.

Crosby, Philip B. *Quality Is Free: The Art of Making Quality Certain*. New York: McGraw-Hill, 1979.

Crosby, Philip B. *Quality without Tears: The Art of Hassle-Free Management*. New York: McGraw-Hill, 1984.

Currid, Cheryl and Company. *Computing Strategies for Reengineering Your Organization*. Rocklin, Calif.: Prima Publishing, 1994.

Davenport, Thomas H. *Process Innovation: Reengineering Work Through Information Technology*. Boston, Mass.: Harvard Business School Press, 1993.

Deming, W. Edwards. *Out of the Crisis*. Cambridge, Mass.: Center for Advanced Engineering Study, Massachusetts Institute of Technology, 1982.

Deming, W. Edwards. *Out of the Crisis*. Cambridge, Mass.: Center for Advanced Engineering Study, Massachusetts Institute of Technology, 1985.

Ealey, Lance A. *Quality by Design: Taguchi Methods and U.S. Industry*. Dearborn, Mich.: ASI Press, 1988.

Edosomwan, J.A. "A Methodology for Comprehensive Productivity Planning." *Productivity Management Frontiers I,* The Netherlands: Elsevier, February 1987.

Edosomwan, J.A. "A Technology-Oriented Total Productivity Measurement Model." *Productivity Management Frontiers I*. The Netherlands: Elsevier, February 1987.

Edosomwan, J.A. "Ergonomic Issues in Computer-Aided Manufacturing." *Trends in Ergonomics IV*. Amsterdam: North-Holland, 1987.

Edosomwan, J.A. *Integrating Productivity and Quality Management*. New York: Marcel Dekker, 1987.

Edosomwan, J.A. "The Impact of a Production-Oriented Quality Circle on Total Productivity." *Productivity Management Frontiers I*. The Netherlands: Elsevier, February 1987.

Edosomwan, J.A. "A Project Management Case Study on Robotic Device Application in a Production Environment." Paper included in *Managing High Technology Projects*. New York: John Wiley & Sons, 1988.

Edosomwan, J.A. *Productivity and Quality Improvement*. England: IFS Publications, 1988.

Edosomwan, J.A. "A Framework for Balancing Productivity and Quality Requirements in Organizations." *Productivity Management Frontiers II*. Geneva, Switzerland: Inderscience, February 1989.

Edosomwan, J.A. *Integrating Innovation and Technology Management: Handbook for Professionals*. New York: John Wiley, 1989.

Edosomwan, J.A. "Planning and Implementing Change in a Manufacturing Organization." *People and Product Management in Manufacturing*. The Netherlands: Elsevier, 1989.

Edosomwan, J.A. "Productivity Research: Enhancing Cooperation Between Industry and Productivity Research Centers." *Productivity Management Frontiers II*. Geneva, Switzerland: Inderscience, February 1989.

Edosomwan, J.A. *A Ten Step Approach for Implementing Total Quality Management*. Fairfax, Va.: Excellence Publications, 1991.

Edosomwan, J.A. *Communication and Team Interaction for Continuous Quality Improvement*. Fairfax, Va.: Excellence Publications, 1991.

Edosomwan, J.A. *Competing Through Quality and Reliability*. Fairfax, Va.: Excellence Publications, 1991.

Edosomwan, J.A. *Competitiveness Through Service Quality*. Fairfax, Va.: Excellence Publications, 1991.

Edosomwan, J.A. *Continuous Improvement Tools and Techniques: A Practical Guide for Improving Quality, Productivity and Performance in Organizations*. Fairfax, Va.: Excellence Publications, 1991.

Edosomwan, J.A. *Continuous Improvement Tools and Techniques—Action Notebook*. Fairfax, Va.: Excellence Publications, 1991.

Edosomwan, J.A. (editor). *Discovering Engineering Futures: Projects and Programs to Support Development of Women, Minorities and Disabled in Engineering*. Atlanta, Ga.: IIE Task Force on W.M.D., 1991.

Edosomwan, J.A. *Project Management in a Total Quality Environment*. Fairfax, Va.: Excellence Publications, 1991.

Edosomwan, J.A. *Quality Through a Self-Directed Workforce*. Fairfax, Va.: Excellence Publications, 1991.

Edosomwan, J.A. *The Winning Quality Manager*. Fairfax, Va.: Excellence Publications, 1991.

Edosomwan, J.A. *Understanding & Implementing Continuous Quality Improvement*. Fairfax, Va.: Excellence Publications, 1991.

Edosomwan, J.A. *Understanding and Implementing TQM*. Fairfax, Va.: Excellence Publications, 1991.

Edosomwan, J.A. *Benchmarking for Continuous Quality Improvement*. Fairfax, Va.: Excellence Publications, 1992.

Edosomwan, J.A. *Building Quality & Productivity in Workteams*. Fairfax, Va.: Excellence Publications, 1992.

Edosomwan, J.A. *Continuous Process Improvement*. Fairfax, Va.: Excellence Publications, 1992.

Edosomwan, J.A. *Continuous Quality Improvement Tools for Purchasing Professionals*. Fairfax, Va.: Excellence Publications, 1992.

Edosomwan, J.A. *Customer and Market-Driven Quality Management*. Milwaukee, Wis.: ASQC Quality Press, 1992.

Edosomwan, J.A. *Implementing Total Quality Management*. Fairfax, Va.: Excellence Publications, 1992.

Edosomwan, J.A. *Improving Total Quality and Productivity in Organizations*. Fairfax, Va.: Excellence Publications, 1992.

Edosomwan, J.A. *Managing Customer Satisfaction*. Fairfax, Va.: Excellence Publications, 1992.

Edosomwan, J.A. *Using Quality Systems to Make and Manage Public Policy*. Fairfax, Va.: Excellence Publications, 1992.

Edosomwan, J.A. *Business Process Reengineering and Performance Improvement*. Fairfax, Va.: Excellence Publications, 1993.

Edosomwan, J.A. *Redesigning and Optimizing Work Processes, Training—Action Notebook*. Fairfax, Va.: Excellence Publications, 1993.

Edosomwan, J.A. *Understanding and Implementing ISO 9000*. Fairfax, Va.: Excellence Publications, 1993.

Edosomwan, J.A. *Market-Driven Quality, Productivity and Total Customer Satisfaction*. Milwaukee, Wis.: ASQC Quality Press, 1994.

Edosomwan, J.A. (editor) and A. Ballakur (associate editor). *Improving Productivity and Quality in Electronics Assembly*. Atlanta, Ga.: IE and Management Press and New York: McGraw-Hill, 1988.

Edosomwan, J.A. and T.M. Khalil (editors). *Technology Management I.* Geneva, Switzerland: Inderscience, 1988.

Edosomwan, J.A., D.J. Sumanth, D.S. Sink and W. Werther (editors). *Productivity Management Frontier III.* Atlanta, Ga.: IE and Management Press, 1991.

Ernst & Young Quality Improvement Consulting Group. *Total Quality: An Executive's Guide for the 1990's.* Homewood, Ill.: Dow Jones-Irwin/ APICS Series in Production Management, 1990.

Feigenbaum, Armand V. *Total Quality Control.* New York: McGraw-Hill, 1983.

Fombrun, Charles J. *Turning Points: Creating Strategic Change in Corporations.* New York: McGraw-Hill, 1992.

Fukuda, Ryuji. *Managerial Engineering: Techniques for Improving Quality and Productivity in the Workplace.* Stamford, Conn.: Productivity Press, 1984.

Garvin, David A. *Managing Quality: The Strategic and Competitive Edge.* New York: The Free Press, 1988.

Gitlow, Howard and Shelly Gitlow. *The Deming Guide to Quality and Competitive Position.* Englewood Cliffs, N.J.: Prentice-Hall, 1986.

Goldston, Mark R. *The Turnaround Prescription: Repositioning Troubled Companies.* New York: Maxwell MacMillan International, 1992.

Gordon, Thomas. *Leader Effectiveness Training.* New York: Bantam Books, 1977.

Groocock, John M. *The Chain of Quality.* Milwaukee, Wis.: ASQC Quality Press, 1986.

Hall, Robert W. *Zero Inventories.* Homewood, Ill.: Dow Jones-Irwin, 1983.

Hammer, Michael and James Champy. *Reengineering the Corporation.* New York: Harper Business, 1993.

Harrigan, Kathryn Rudie. *Managing Maturing Businesses: Restructuring Declining Industries and Revitalizing Troubled Operations.* Lexington, Mass.: Lexington Books, 1988.

Harrington, H. James. *Excellence—The IBM Way.* Milwaukee, Wis.: ASQC Quality Press, 1986.

Harrington, H. James. *The Quality/Profit Connection.* Milwaukee, Wis.: ASQC Quality Press, 1986.

Harrington, H. James. *The Improvement Process—How America's Leading Companies Improve Quality.* New York: McGraw-Hill, 1987.

Harrington, H. James. *Excellence—The IBM Way.* Milwaukee, Wis.: Publisher's Quality Press, 1988.

Harrington, H. James. *Business Process Improvement: The Breakthrough Strategy for Total Quality, Productivity and Competitiveness.* New York: McGraw-Hill, 1991.

Hayse, Robert H. and Steven C. Wheelwright. *Restoring Our Competitive Edge: Competing Through Manufacturing.* New York: John Wiley & Sons, 1984.

Herzberg, Frederick. *Work and the Nature of Man.* Cleveland, Ohio: Cleveland World, 1966.

Hickman, Craig R. and Michael A. Silva. *Creating Excellence.* New York: New American Library, 1984.

Hunt, V. Daniel. *Reengineering.* Essex Junction, Vt.: Oliver Wight Publications, 1993.

Imai, Masaaki. *Kaizen: The Key to Japan's Competitive Success.* New York: Random House, 1986.

Ishikawa, Kaoru. *Guide to Quality Control.* White Plains, N.Y.: Kraus International, 1982.

Ishikawa, Kaoru. *Guide to Quality Control.* Japan: Asian Productivity Organization, 1984.

Ishikawa, Kaoru. *What Is Total Quality Control? The Japanese Way.* Englewood Cliffs, N.J.: Prentice-Hall, 1985.

Johansson, Henry J., Patrick McHugh, A. John Pendleburg and William A. Wheeler III. *Business Process Reengineering.* Chichester, England: John Wiley & Sons, 1993.

Juran, J.M. *Managerial Breakthrough.* New York: McGraw-Hill, 1964.

Juran, J.M. *Quality Control Handbook.* New York: McGraw-Hill, 1974.

Juran, J.M. *Juran on Leadership for Quality.* Milwaukee, Wis.: ASQC Quality Press, 1986.

Juran, J.M. *Juran on Planning for Quality.* New York: The Free Press, 1988.

Juran, J.M. *Juran's Quality Control Handbook.* New York: McGraw-Hill, 1988.

Juran, J.M. *Juran on Leadership for Quality: An Executive Handbook*. New York: The Free Press, 1989.

Juran, J.M. and Frank M. Gryna, Jr. *Quality Planning and Analysis*. New York: McGraw-Hill, 1980.

Kanter, Rosebeth Moss. *The Change Masters*. New York: Simon & Schuster, 1983.

Kepner, Charles H. and Benjamin B. Tregoe. *The New Rational Manager*. Princeton, N.J.: Princeton Research Press, 1981.

Kume, Hitoshi. *Statistical Methods for Quality Improvement*. New York: The Association for Overseas Technical Scholarship, UNIPUB, 1985.

Laurence, Peter J. *The Peter Principle*. New York: Morrow, 1969.

Likert, Rensis. *New Patterns of Management*. New York: McGraw-Hill, 1961.

Mann, Nancy R. *The Keys to Excellence*. Santa Monica, Calif.: Prestwick Books, 1985.

Mansir, Brian E. and Nicholas R. Schacht. *Introduction to the Continuous Improvement Process: Principles and Practices*. LMI Report IR806R1, August 1989.

McGregor, Douglas. *Leadership and Motivation*. Cambridge, Mass.: The MIT Press, 1983.

McGregor, Douglas. *The Human Side of Enterprise*. New York: McGraw-Hill, 1985.

Michalak, Donald F. and Edwin G. Yager. *Making the Training Process Work*. New York: Harper & Row, 1979.

Mitzenberg, H. *The Nature of Managerial Work*. New York: Harper & Row, 1973.

Monden, Yasuhiro. *Toyota Production System*. Norcross, Ga.: Institute of Industrial Engineers, 1982.

Morris, Daniel and Joel Brandon. *Reengineering Your Business*. New York: McGraw-Hill, 1993.

Nadler, Gerald and Shozo Hibino. *Breakthrough Thinking: Why We Must Change the Way We Solve Problems, and the Seven Principles to Achieve This*. Rocklin, Calif.: Prima Publishing, 1990.

Nayak, P. Raganath and John M. Keeteringham. *Breakthroughs!* San Diego, Calif.: Pfeiffer & Company, 1994.

Ouchi, William G. *Theory Z*. Reading, Mass.: Addison-Wesley, 1981.

Peters, Tom. *Thriving on Chaos*. New York: Alfred A. Knopf, 1987.

Porter, Michael E. *Competitive Strategy*. New York: The Free Press, 1984.

Schein, Edgar. *Organizational Culture and Leadership*. San Francisco, Calif.: Jossey-Bass, 1985.

Scherkenbach, William W. *The Deming Route to Quality and Productivity*. Washington, D.C.: CEE Press Books, George Washington University, 1986.

Scherkenbach, William W. *The Deming Route to Quality and Productivity*. Rockville, Md.: Mercury Press, 1988.

Scholtes, Peter R. et al. *The Team Handbook—How to Use Teams to Improve Quality*. Madison, Wis.: Joiner Associates, 1988.

Schonberger, Richard J. *Japanese Manufacturing Techniques: Nine Hidden Lessons in Simplicity*. New York: The Free Press, 1982.

Schonberger, Richard J. *World Class Manufacturing—The Lessons of Simplicity Applied*. New York: The Free Press, 1986.

Shores, Richard A. *Survival of the Fittest*. Milwaukee, Wis.: ASQC Quality Press, 1986.

Spendolini, Michael J. *The Benchmarking Book*. New York: AMACOM, 1992.

Stalk, George and Thomas M. Hout. *Competing Against Time: How Time-Based Competition Is Reshaping Global Markets*. New York: The Free Press, 1990.

Stratton, A. Donald. *An Approach to Quality Improvement that Works*. Milwaukee, Wis.: ASQC Quality Press, 1986.

Strebel, Paul. *Breakpoints, How Managers Exploit Radical Business Change*. Boston, Mass.: Harvard Business School Press, 1992.

Taguchi, Genichi and Yuin Wu. *Introduction to Off-Line Quality Control*. Tokyo, Japan: Central Japan Quality Control Association, 1979.

Tichy, Noel M. and Stratford Sherman. *Control Your Destiny or Someone Else Will*. New York: Doubleday, 1993.

Tomasko, Robert M. *Rethinking the Corporation*. New York: AMACOM, 1993.

Townsend, Patrick L. *Commit to Quality*. New York: John Wiley & Sons, 1986.

Tribus, Myron. *Quality First: Selected Papers on Quality and Productivity Improvement.* Washington, D.C.: American Quality and Productivity Institute, National Society of Professional Engineers, March 1988.

Tushman, Michael L. and William L. Moore. *Readings in the Management of Innovation.* New York: Ballinger, 1988.

Wadsworth, Stephens and Blan Godfrey. *Modern Methods for Quality Control and Improvement.* New York: John Wiley & Sons, 1986.

Walton, Mary. *The Deming Management Method.* New York: Dodd, Mead & Company, 1986.

Walton, Mary. *Deming Management at Work.* New York: Putnam, 1990.

Warton, Richard S. *Up and Running.* Boston, Mass.: Harvard Business School Press, 1989.

Watson, Gregory. *The Benchmarking Workbook: Adapting Best Practices for Performance Improvement.* Cambridge, Mass.: Productivity Press, 1992.

Weiss, Alan. *Making It Work: Turning Strategy into Action Throughout Your Organization.* New York: Harper Collins, 1990.

Western Electric. *Statistical Quality Control Handbook.* Easton, Pa.: Mack Publishing, 1977.

Wheeler, Donald J. and David S. Chambers. *Understanding Statistical Process Control.* Knoxville, Tenn.: Statistical Process Controls, 1986.

Whitney, John O. *Taking Charge: Management Guide to Troubled Companies and Turnarounds.* Homewood, Ill.: Business One Irwin, 1987.

Zemke, Ron and Dick Schaaf. *The Service Edge: 101 Companies that Profit from Customer Care.* New York: New American Library, 1989.

Zuboff, Shoshana. *In the Age of the Smart Machine: The Future of Work and Power.* New York: Basic Books, 1988.

Articles

Edosomwan, J.A. "A Conceptual Framework for Productivity Planning." *Industrial Engineering.* January 1986.

Edosomwan, J.A. "Managing Technology in the Workplace—A Challenge for Industrial Engineers." *Industrial Engineering.* February 1986.

Edosomwan, J.A. "A Program for Managing Productivity and Quality." *Industrial Engineering.* January 1987.

Edosomwan, J.A. "Industrial Engineer Roles Are Vaned and Numerous in High Technology Environment." *Industrial Engineering.* December 1987.

Edosomwan, J.A. "Managing Productivity and Quality—A Challenge for Industrial Managers." *Industrial Management.* September/October 1987.

Edosomwan, J.A. "Ten Design Rules for Knowledge Based Expert Systems." *Industrial Engineering.* August 1987.

Edosomwan, J.A. "Managing Change in a Market-Driven Enterprise." *Industrial Management.* September/October 1989.

Edosomwan, J.A. "Preparing the Professionals' Skills for the Factory of the Future." *Industrial Engineering.* November 1989.

Edosomwan, J.A. "Assessing Your Organization's TQM Posture and Readiness to Successfully Compete for the Malcolm Baldrige Award." *Industrial Engineering.* 1991.

Edosomwan, J.A. "Five Initiatives for Your Customer Satisfaction Level." *The Quality Observer.* December 1991.

Edosomwan, J.A. "Implementing Customer Driven Quality Management." *The Quality Observer.* November 1991.

Edosomwan, J.A. "Malcolm Baldrige National Quality Award: Focus on Total Customer Satisfaction as a Key Element in Winning." *Industrial Engineering.* July 1991.

Edosomwan, J.A. "Adopting a Customer First Philosophy." *The Quality Observer.* April 1992.

Edosomwan, J.A. "A Passion for Customer Satisfaction." *The Quality Observer.* March 1992.

Edosomwan, J.A. "Balancing Productivity and Quality Results." *The Quality Observer.* December 1992.

Edosomwan, J.A. "Customer Satisfaction Through Continuous Process Improvement." *The Quality Observer.* May 1992.

Edosomwan, J.A. "Dealing with Customer Complaints." *The Quality Observer.* February 1992.

Edosomwan, J.A. "Implementation Strategies for Quality Program." *Industrial Engineering.* October 1992.

Edosomwan, J.A. "Leadership for Continuous Quality Improvement." *The Quality Observer.* August 1992.

Edosomwan, J.A. "On Becoming a Customer-Driven Organization." *The Quality Observer.* January 1992.

Edosomwan, J.A. "Preparing the Workforce to Accept Changes Required for Quality Improvement." *The Quality Observer.* October 1992.

Edosomwan, J.A. "Seven Commandments for Achieving Total Customer Satisfaction." *The Quality Observer.* June 1992.

Edosomwan, J.A. "Handling Obstacles to Organizational Process Reengineering and Total Quality Management Implementation." *The Quality Observer.* January 1994.

Edosomwan, J.A. and L. Forbes. "U.S. Engineering Manpower Shortages—A Collision Course with Crisis." *Industrial Engineering.* September 1990.

Edosomwan, J.A. and T.M. Khalil. "Slips and Falls: A Comprehensive Approach to Accident Prevention." *Journal of Professional Safety.* June 1981.

Edosomwan, J.A. and C. Marsh. "Streamlining Material Flow for Just-in-Time Production." *Industrial Engineering.* January 1991.

Edosomwan, J.A. and W. Savage-Moore. "The Making of Women and Minorities in Engineering." *IIE Focus.* October 1990.

Edosomwan, J.A. and W. Savage-Moore. "The Malcolm Baldrige Award: Could Your Organization Win One?" *Industrial Engineering.* February 1991.

Farrow, John. "Quality Audits: An Invitation to Managers." *Quality Progress.* January 1987.

Fuller, F. Timothy. "Eliminating Complexity from Work: Improving Productivity by Enhancing Quality." *National Productivity Review.* Autumn 1985.

Garvin, David A. "Quality on the Line." *Harvard Business Review.* September–October 1983.

Hayse, Robert H. "Why Japanese Factories Work." *Harvard Business Review.* July–August 1981.

Joiner, Brian and Peter Scholtes. "The Quality Manager's New Job." *Quality Progress.* October 1986.

Juran, J.M. "The Quality Trilogy." *Quality Progress.* August 1986.

Kacker, Raghu N. "Quality Planning for Service Industries." *Quality Progress.* August 1988.

Kaplan, Robert S. "Yesterday's Accounting Undermines Production." *Harvard Business Review*. July–August 1984.

Kaplan, Robert S. "Measuring Manufacturing Performance: A New Challenge for Managerial Accounting Research." *The Accounting Review*. February 1985.

Leonard, Frank S. and W. Earl Sasser. "The Incline of Quality." *Harvard Business Review*. September–October 1982.

McLean, Gary N. and Sam Parkenham-Walsh. "An In-Process Model for Improving Quality Management Processes." *Consultation*. Volume 6, Number 3, Fall 1987.

Melan, Eugene H. "Process Management in Service and Administrative Operations." *Quality Progress*. June 1985.

Miller, Jeffery G. and Thomas E. Vollmann. "The Hidden Factory." *Harvard Business Review*. September–October 1985.

Moen, Ronald D. and Thomas W. Nolan. "Process Improvement." *Quality Progress*. September 1987.

Scholtes, Peter R. and Heero Harquebord. "Six Strategies for Beginning the Quality Transformation, Part I." *Quality Progress*. July 1988.

Scholtes, Peter R. and Heero Harquebord. "Six Strategies for Beginning the Quality Transformation, Part II." *Quality Progress*. August 1988.

Schonberger, Richard J. "A Revolutionary Way to Streamline the Factory." *The Wall Street Journal*. November 15, 1982.

Schonberger, Richard J. "Integration of Cellular Manufacturing with Just-in-Time Production." *Industrial Engineering*. December 1982.

Schonberger, Richard J. "Production Workers Bear Major Responsibility in Japanese Industry." *Industrial Engineering*. December 1982.

Schonberger, Richard J. "Rationalizing the Workplace: First Step in Implementing Robotics." *Robot-X News*. February 15, 1983.

Schonberger, Richard J. "Just-in-Time Purchasing: A Challenge for U.S. Industry." *California Management Review*. Autumn 1983.

Schonberger, Richard J. "Work Improvement Programs: Quality Circles Compared with Traditional and Evolving Western Approaches." *International Journal of Operations and Production Management*. Volume 3, Number 2, 1983.

Schonberger, Richard J. "The Quality Dividend of Just-in-Time Production." *Quality Progress*. October 1984.

Schonberger, Richard J. and Marc Schniederjans. "Reinventing Inventory Control." *Interfaces*. May–June 1984.

Schultz, Louis E. "Creating a Vision for Strategy and Quality: A Way to Help Management Assume Leadership." *Concepts in Quality Proceedings*. November 1988.

Shimoyamada, Kaoru. "The President's Audit: QC Audit at Komatsu." *Quality Progress*. January 1987.

Sullivan, Laurance P. "Quality Function Deployment." *Quality Progress*. June 1986.

Suzaki, Kiyoshi. "Work-in-Process Management: An Illustrated Guide to Productivity Management." *Production and Inventory Management*. Third Quarter 1985.

Walleigh, Richard C. "What's Your Excuse for Not Using JIT?" *Harvard Business Review*. March–April 1986.

Acknowledgment of Technical Sources

The author wishes to recognize all of the authors and publications that were referenced in this book.

- The source of the LMI CIP Transformation Model is *An Introduction to the Continuous Improvement Process: Principles and Practices* by Brian E. Mansir and Nicholas R. Schacht, Logistics Management Institute (1989).

- The source of the DSMC Q&PMP Transformation Model is *Quality and Productivity Management Practices on Defense Programs,* Fort Belvoir, Va.: Defense Systems Management College (1988).

- The source of the DSMC/ATI Performance Improvement Model is *Quality and Productivity Management Practices on Defense Programs,* Fort Belvoir, Va.: Defense Systems Management College (1988).

- The source of the PMI Leadership Expectation Setting Model is *L.E.S. Management* by Louis E. Schultz, Bloomington, Minn.: Process Management Institute (1989).

- The source of the Moen and Nolan Strategy for Process Improvement is "Process Improvement: A Step-by-Step Approach to Analyzing and Improving a Process" by Ronald D. Moen and Thomas W. Nolan, *Quality Progress* (Sept. 1987).

- The source of the Quality Journal Model is *The Quality Journal* by Louis E. Schultz, Bloomington, Minn.: Process Management Institute (1989).

- The source of the LMI CIP Personal Improvement Model is *Introduction to the Continuous Improvement Process: Principles and Practices* by Brian E. Mansir and Nicholas R. Schacht, Logistics Management Institute (1989).

- The source of the NPRDC Process Improvement Model is *Defining the Deming Cycle: A Total Quality Management Process Improvement Model* by S.L. Dockstader and A. Houston, San Diego, Calif.: Navy Personnel Research and Development Center (1988).

- The source of the FPL Quality Improvement Story is the Florida Power and Light Company, 700 Universe Boulevard, Juno Beach, Florida 33408.

- The source of the Joiner Associates' Model of Progress is *The Team Handbook* by Peter R. Scholtes et al., Madison, Wis.: Joiner Associates, Inc. (1988).

- The source of the LMI CIP Process Improvement Model is *Introduction to the Continuous Improvement Process: Principles and Practices* by Brian E. Mansir and Nicholas R. Schacht, Logistics Management Institute (1989).

- The source of the work force involvement suggestions is *Introduction to the Continuous Improvement Process: Principles and Practices* by Brian E. Mansir and Nicholas R. Schacht, Logistics Management Institute (1989).

Acknowledgment of Clients

The author wishes to thank all of the clients of Johnson & Johnson Associates, Inc. (JJA) who have provided environments which enabled the testing of some of the contents of this book. The following is a partial list of the organizations that have either participated in workshops or obtained consulting services from JJA.

U.S. Public Health
Defense Logistics Agency
Department of Energy
Navy Audit Services
U.S. General Accounting Office
Internal Revenue Service
City of St. Louis
Environmental Protection Agency
Census Bureau
Virginia Housing Development
Washington Gas
Institute of Industrial Engineers
The Analytic Sciences Corporation
General Dynamics
Arthur Young
San Jose State University
Strategic Defense Initiative Organization
Syscon
University of Manitoba

Department of Interior
Tandem Computers
Arthur D. Little
Washington, D.C. Public Schools
Delta Research Corporation
USM Corporation
Nunes, Scholefield, DeLeon & Co.
U.S. Coast Guard
Monterey Water Pollution Control Agency
U.S. Mint
Montgomery Elevator Company
National Credit Union
Xilinx
IBM
NASP
Aesculap
PHP, Inc.
Lockheed
BTG, Inc.

PRC, Inc.
Citibank
Synetex
Medasonics
TMS
Carson Consulting
Veda
Amdahl
DC Electronics
Solitron
New York Metropolitan Transit
    Authority
City of Fairfax
New York State Governor's
    Office
George Mason University
Marriott Corporation
PacTel Meridian Systems
Arthur Anderson
St. Louis Airport Authority

Scientific Technologies, Inc.
Xerox Engineering Systems
NAPM
American Society for Quality
    Control
Howard University
Diversified Internatscience
Pacific Gas & Electric
Fairfax Opportunities Unlimited
Hewlett-Packard
Santa Cruz Operations
Herra Company
American Express
Northern Telecom
Solitron Devices, Inc.
Defense Mapping Agency
American Productivity and
    Inventory Control Society
RMS Technologies
Bureau of Engraving and
    Printing

Index

Selected Courses on Organizational and Process Reengineering

The following courses on organizational and process reengineering are offered by Johnson & Johnson Associates, Inc.

- Reengineering Organizational Work Process
- Understanding Reengineering Tools and Techniques
- Facilitating Reinvention and Reengineering Teams
- Managing the Reengineering Process
- Leadership for Process Reengineering
- Reinventing Government with Process Reengineering
- Benchmarking for Process Reengineering
- Organizational Transformation Principles
- Leading and Managing Change
- Optimizing People, Technology and Process Changes
- Redesigning Organizational Structures
- Career Continuation and Transition Planning
- Facilitating Change Management Teams
- Understanding and Minimizing Resistance to Change
- Managing Displaced Workers
- Reinventing Government Organizations
- Successful Job Redesign and Enrichment
- Tools and Techniques for Change Management
- Managing Incremental and Radical Change Processes
- Personal and Professional Transition Management
- Communication and Transition Planning

For additional information call (703) 359-5969